NEW TECHNOLOGIES FOR EMISSION CONTROL IN MARINE DIESEL ENGINES

NEW TECHNOLOGIES FOR EMISSION CONTROL IN MARINE DIESEL ENGINES

MASAAKI OKUBO

TAKUYA KUWAHARA

ELSEVIER

Butterworth-Heinemann
An imprint of Elsevier

Butterworth-Heinemann is an imprint of Elsevier
The Boulevard, Langford Lane, Kidlington, Oxford OX5 1GB, United Kingdom
50 Hampshire Street, 5th Floor, Cambridge, MA 02139, United States

Library of Congress Cataloging-in-Publication Data
A catalog record for this book is available from the Library of Congress

British Library Cataloguing-in-Publication Data
A catalogue record for this book is available from the British Library

ISBN: 978-0-12-812307-2

For information on all Butterworth-Heinemann publications visit our website at https://www.elsevier.com/books-and-journals

Publisher: Matthew Deans
Acquisition Editor: Carrie Bolger
Editorial Project Manager: Carrie Bolger
Production Project Manager: Sruthi Satheesh
Cover Designer: Matthew Limbert

Typeset by TNQ Technologies

Contents

Preface

Recently, emission controls for combustion engines have become an important topic in environmental technology. Emission controls can be classified into two types: combustion improvement for low-emission engines and aftertreatment/exhaust gas cleaning. The main objective of this book is to present readers with an easy-to-understand resource for the foundation and application of technologies for marine diesel engine emission control. This book describes different types of emission control technologies for marine diesel engines and attempts to be the first book of its type to treat these topics. Further, this book provides a bridge from fundamentals to practical industry applications in environmental engineering. Diesel engines have the advantages of low carbon dioxide (CO_2) emission, compact size, and higher fuel efficiency relative to output power. However, their emissions contain harmful particulate matter (PM) and nitrogen oxide (NO_x), and control technologies for them are required. Combustion improvements for low-emission engines are important, such as fuel injection systems and dual-fuel systems, and, therefore, should be investigated primarily as emission control technologies. On the other hand, aftertreatment technologies for emissions have been extensively developed in recent years and are the most effective means of emission control. Because regulations on automobile emission are already in place in developed countries, marine diesel engines should be the next target. The NO_x concentration in marine diesel emissions is relatively high (500−1000 ppm). Furthermore, the marine diesel emissions contain sulfur oxides (SO_x) and PM. This is because the engines are operated using marine diesel oil or marine fuel oil fuels, which contain a significant amount of sulfur with the concentration of several percent at most. It has been suggested that emissions from marine diesel engines of ocean-faring ships could contribute to climate change and increase the risk/incidence of respiratory diseases; therefore, stringent regulations and its keeping are needed.

This book mainly focuses on emission control for marine diesel engines, but emission controls for automobile and off-road engines are also described. Because marine diesel engines are usually fueled by heavy oil including sulfur and nitrogen, emission control is more difficult to incorporate compared with automobile diesel engines. This is because the latter utilizes gas oil, which contains much less sulfur and nitrogen. However, because the rotational speeds of marine diesel engines are usually steady, emissions are induced at a steady rate and thus are easier to control. Conversely, automobile engine emissions fluctuate because the rotational speed of the engine varies more frequently; thus, its emissions are more difficult to control. The complete removal of these pollutants from engines in our laboratory was very exciting experience for us, and we hope that some of our fascinations with this technology is expressed in the book.

Our initial studies were motivated by the desire to clean stationary fossil fuel (coal-, oil-, and gas-fired) power generation plants emissions and clean exhaust gases, including dioxins, emitted by garbage incinerators. The cleaning methods utilized new technologies that incorporated nonthermal plasmas. Recently, almost all power generation plants in Japan have been equipped with an environmental system composed of selective catalytic reduction, electrostatic precipitator, and soda-lime gypsum methods, and simultaneous removal of NO_x−PM−SO_x with high efficiency has been achieved. As a result, Japanese skies have been kept very clear. However, early in the 2000s, emission control for automobile diesel engines became a problem in Japan and in other western countries. We began JST (Japan Science and Technology Agency) project to control diesel engine emissions. The main cleaning technology used in the project was also nonthermal plasma. Similar to the outcome of other diesel engine projects, our project developed a new aftertreatment system for super-clean automobile diesel engines. As a result of another our combustion emission cleaning project, a super-clean industrial boiler was successfully developed with a NO_x emission level of less than 10 ppm. Recently, marine diesel engines with NO_x, SO_x, and PM controls have attracted attention, and efforts have been devoted to further develop this research field in marine engineering. We believe that these research developments should be documented, and thus we have written this book to convey technologies to future generations.

These projects were accomplished mainly at Osaka Prefecture University in Sakai city, Osaka Prefecture, Japan, which has been the main research institution of the authors. Sakai city has prospered ocean trade since long ago. In the 16th century, it was called "Oriental Venice" known as "Saccai" to Europeans and was imaged as a prosperous international trade port. As an image of the prosperous Sakai of those days, there is a drawing picture published in a book written by Dutch missionary and scholar, Arnoldus Montanus, in the 17th century as shown in Fig. A.1 in A.2

of Appendix. There are various international ships going at Sakai port. At the present time, Sakai city has developed as an industrial city in Japan with a large factory zone in the coastal area. We are delighted to be able to publish this technical book on the marine environmental technology from Osaka Prefecture University located in the traditional Japanese city.

The recent developments in aftertreatment technologies and their fundamental aspects are included in this book. Selected applications of emission control technologies are also described. Some of them have been completed in the commercial stage, while others are still in the early stage of development. The technical details and test results are described based on the fundamental knowledge of environmental protection engineering. One of the authors, Masaaki Okubo, has experience teaching master's courses of graduate school at Osaka Prefecture University. The title of one of the classes is Environmental Protection Engineering for the Department of Mechanical Engineering and the Department of Aerospace and Marine Engineering. This book is intended to be a textbook for the class. Although the book treats mainly emission control in marine diesel engines, it includes descriptions of the overall technology for air cleaning. Discussions with Dr. Tomoyuki Kuroki at Osaka Prefecture University have been enthusiastic and productive, and we are appreciative of his discussion with us on the various topics treated in this book.

Emission control technology is an interdisciplinary topic that is physical and chemical in nature, with applications in mechanical, chemical, environmental, and marine engineering. Because practitioners in these fields have diverse backgrounds, an effort has been made to devise a technology that is sufficiently self-contained to be accessible to engineers, scientists, and students from many fields.

Chapter 1 provides an introduction to classify the different types of ships, fundamentals of marine diesel engines, their mechanism, and emissions. Further, targets of emission control in this book is described. Chapter 2 treats recent marine and automobile emission regulations for NO_x, SO_x, PM, and fuels. Chapter 3 treats the principle and design of emission control systems for marine and automotive applications. The fundamentals of control method for air pollutants generated in exhaust gas such as NO_x, SO_x, and PM are presented. The fundamentals of environmental protection engineering are also described for enhanced understanding of that field. The concept of electrostatics, plasma discharge, and electrostatic precipitation are rarely included in any contemporary curriculum. Accordingly, their important fundamental aspects are included in the chapter. Furthermore, the fundamentals of power electronics for plasma generation are described. Chapter 4 treats operation examples of emission control system for marine diesel engines. The achievements of our plasma super-clean diesel engine project on PM and NO_x reductions are also described. Chapter 5 treats prospects for marine diesel engine emission control. Advanced NO_x, SO_x, and volatile organic compound control technologies; $PM_{2.5}$ emission and health effect; CO_2 emission reduction; past, current, and future market size of aftertreatment systems; and the operation and maintenance of various aftertreatment systems are treated. These topics are explained in this book as self-contained descriptions based on many kinds of literature and with appendix.

In this book, we focus on emission control technologies and therefore cannot provide a complete explanation of the details associated with various engineering fields. When this book is used as a textbook, we would like to inform the teachers in mechanical engineering that we are unable to provide a complete description of ships' structure, mechanism of diesel engines, fluid engineering, and combustion engineering in this book. To sufficiently understand them, it is necessary to refer to textbooks on mechanical and marine engineering. We would like to inform the teachers in electrical engineering that we are unable to provide a complete description of high voltage engineering, electrostatics engineering, and plasma engineering. It is necessary to refer textbooks on electrical and plasma engineering for further study. We would like to inform the teachers in environmental engineering that we are unable to provide a complete description of environmental engineering and a wide range of aftertreatment technologies currently being used. To sufficiently understand them, it is necessary to refer to textbooks on environmental engineering.

The two authors worked closely and firstly wrote an outline of each chapter. Subsequently, we wrote each chapter separately and finally completed the manuscript through discussions to maintain comprehensive unity. The help we received during the creation of this work is appreciated. We thank Ms. Miho Onishi for her sustained efforts in typing the manuscript. It has also been a pleasure to work with the publisher's staff and Ms. Carrie Bolger, who is the editor of the book.

Collaborations with many colleagues over a number of years have been very enriching and enjoyable. We discussed the future prospects of aftertreatment systems with Dr. Keiichiro Yoshida at Osaka Institute of Technology and Dr. Hidekatsu Fujishima at Osaka Prefecture University. Research studies with many students for over 20 years at Osaka Prefecture

University have been enriching and enjoyable. Most are credited in this book through citations of their published work. We have truly enjoyed studying and experimenting with emission control; we have found it to be a very important research topic that has serious potential to improve the lives and health of humanity. We hope that this book will be beneficial not only among marine engineers and students but also among those in other fields such as mechanical, electrical, chemical, and environmental engineering to get fundamental knowledge on technologies for emission control in marine diesel engines.

Masaaki Okubo
Osaka Prefecture University, Sakai, Japan
Takuya Kuwahara
Nippon Institute of Technology, Saitama, Japan

CHAPTER 1

Introduction

Contents

1.1 Classification of ships

A ship is a transportation system and a vessel that travels on water supported by buoyant force. There are many types of ships, such as merchant ships, special operation ships, and other miscellaneous ships; there are also many ideas on how they should be classified. Before the explanation of marine diesel engines and emission controls are presented, an idea of the classification, the contribution of ships, and the status of marine transportations are provided [1–4].

1.1.1 Merchant ships

Products and energy necessary for our clothing, food, lives, and raw materials for the industry are transported by merchant ships. For example, in Japan,

New Technologies for Emission Control in Marine Diesel Engines
ISBN 978-0-12-812307-2
https://doi.org/10.1016/B978-0-12-812307-2.00001-8

which is an island country, the total amount of imported and exported goods exceeds 900 million ton annually, approximately 99.6% of which are carried by merchant ships [1]; less than 0.4% is carried by airplanes. Merchant ships are classified as cargo ships, tankers, product-carrying ships, passenger ships, and other miscellaneous merchant ships. These different classifications are explained below.

Cargo ships

Cargo ships are classified into general cargo ships or freighters, bulk carriers, container ships, and others. General cargo ships transport mainly cargo, goods, and materials from one port to another. A bulk carrier, such as a bulker or dry bulker, is a type of merchant ship that is used to transport large amounts of unpackaged bulk cargo, such as iron ore, coal, cereal, salt, aluminum, and copper ore. Bulk carriers come in many different sizes, and a particular size is utilized depending on the amount of bulk cargo that must be transported and the size of the port that receives it. Fig. 1.1.1 shows a photograph of bulk carrier. The largest type of bulk carrier is called capesize bulker ship. The capesize bulker ship is a type of ship that cannot pass through the Panama Canal because of its large size; therefore, it travels between the Atlantic and Pacific oceans, passing by the Cape of Good

Figure 1.1.1 Bulk carrier ship. *(By David Álvarez López, licensed under CC BY 2.0, https://www.flickr.com/photos/david_a_l/32894529613/.)*

Hope in Africa. The name "capesize" was derived from its required travel path. Another example of a capesize ship is a bulk carrier for coal. This ship carries coal for coal-fired power generation plants and is designed based on the depth-specialized berth of the power generation plant and the moving area of coal-lifting machines. The width is generally extended to increase coal-carrying capacity and tonnage to the 80,000–90,000 class. Fig. 1.1.2 shows a photograph of a container ship. Container ships carry cargo inside overseas shipping containers that meet international regulations. The capacity of a container ship is measured in twenty-foot equivalent units, which are the number of 20-foot containers that a ship can carry, although the majority of containers used today are 40 feet ($\cong$ 12 m) in length.

Tankers

Tankers transport chemical products of liquid and gas. An oil tanker transports liquid products such as crude oil, refined petroleum products such as gasoline, and chemical solutions. An oil tanker is a typical example of this class of ship, which can transport crude oil in bulk. Fig. 1.1.3 shows a photograph of a large crude oil tanker. Oil tankers transport oil from

Figure 1.1.2 Container ship. *(By Ian Kennedy, licensed under CC BY-SA 2.0, https://www.flickr.com/photos/clankennedy/10865084363/.)*

Figure 1.1.3 Crude oil tanker. *(By Frans Berkelaar, licensed under CC BY-SA 2.0, https://www.flickr.com/photos/28169156@N03/19863834578/.)*

countries rich in oil resources. Crude oil tankers carry oil that supports the demands of human life and economic demands. A crude oil tanker, such as the very large crude carrier, is generally utilized, the mass of which is more than 200,000−320,000 ton. Different cargo tanks inside the ships allow for different types of crude oil to be transported separately. Furthermore, liquefied natural gas (LNG) tankers are often observed transporting natural gas because of increased demand as safe and clean energy resources. Fig. 1.1.4 shows a photograph of an LNG tanker. LNG tankers have tanks that have functions similar to a vacuum bottle. A typical component of natural gas is methane. If it is transported in the gas phase, a large volume is required for its storage. However, the volume is reduced to less than 1/600 by liquefaction, which enables high-efficiency transport. Because the boiling point of LNG is very low, −161.5°C, tanks are made with specialized steel such as nickel, stainless steel, or aluminum alloy. For LNG tankers, steam turbine engines are often used because the gasified LNG in tanks can be used as fuel during transportation. There are other types of tankers as well.

Figure 1.1.4 Liquefied natural gas tanker. *(By Ken Hodge, licensed under CC BY 2.0, https://www.flickr.com/photos/40132991@N07/7501108978/.)*

A product tanker carries gasoline, naphtha (low-quality gasoline), kerosene, gas oil (GO), and other oil products. A chemical tanker mainly carries liquid chemical products such as benzene, toluene, and alcohols. A liquefied petroleum gas (LPG) tanker carries LPG. Some LPG tankers are also multipurpose ships that can carry liquefied ammonia, which is the raw material of textiles and fertilizers.

Product-carrying ships

There are also ships that can carry processed products, which are different from bulkers that carry natural resources and tankers that carry energy resources. There are many types of product-carrying ships. One type is a container ship that carries apparel and various types of household goods. Reefer refrigerator ships are used to transport fresh goods such as fruits, vegetables, meat, and seafood. Ships carrying automobiles are usually called pure car carriers or pure car and truck carriers and are specifically designed to carry automobiles and construction machines. Fig. 1.1.5 shows a

Figure 1.1.5 Automobile-carrying ship. *(Photo by courtesy of Mitsui O.S.K. Lines.)*

photograph of an automobile-carrying ship with a capacity of 6800 cars. The ship usually does not have loading facilities such as cranes; instead, the automobiles are driven by humans, who move the cars from the quay to the interior of the ship [3]. This type of automobile loading method is called RO/RO (roll-on/roll-off), and these types of ships are sometimes called RO/RO ships.

Passenger ships

There are two types of passenger ships. One type is a ferry, which can periodically transport passengers, automobiles, and freight cars such as trucks and semitrailers. For example, one Japanese ferry can carry 632 passengers, 77 automobiles, and 154 trucks. Generally, there are parking spaces for automobiles and freight in the lower region of the ship interior. Automobiles are moved from the ship to the port through ramps over the deck using the RO/RO method. Passenger rooms and restaurants are located above the deck. The other type of passenger ship is a cruise ship. Fig. 1.1.6 shows a photograph of a passenger cruise ship used for travel and entertainment. The ship will arrive at the harbor slowly as passengers are dining, watching a movie/show, or gazing at the ocean.

Figure 1.1.6 Cruise passenger ship, Asuka II. *(By Kabacchi, licensed under CC BY 2.0, https://www.flickr.com/photos/kabacchi/3591076067/.)*

Other miscellaneous merchant ships

Miscellaneous ship types exist as well. For example, a barge is a flat-bottomed ship built to transport heavy cargo through inland waterways, such as rivers and canals, and in ports. Many of them are not equipped with engines and cannot sail by themselves.

1.1.2 Special operation ships

Special operation ships are roughly classified into (1) operation ships, (2) helper ships, and (3) other special operation ships. These different types are explained below.

Operation ships

There are many types of operation ships. Fig. 1.1.7 shows a photograph of a type of operation ship that sets, repairs, and collects submarine cables for telecommunication [3]. Because it is necessary to keep the ship in a fixed position while this work is performed, it has a dynamic positioning system that holds the ship in place against the effects of displacing forces such as waves and wind.

Figure 1.1.7 Operation ship that sets, repairs, and collects submarine cables. *(Photo by courtesy of Kokusai Cable Ship Co., Ltd.)*

Helper ships

A tugboat is a helper ship that pushes and pulls ships and structures over water, whose size varies depending on the application. The 150–200-ton class is typically used to assist large ships in the harbor, and the boat assists in detaching from it. The several thousand-ton class tugboat is used to transport large ships and structures over water. Because it is necessary to move a ship or structure much larger than the ship itself, it is equipped with a powerful engine.

Other special operation ships

Other types of special operation ships exist. A heavy-duty ship is one that specializes in carrying heavy objects, such as manufacturing plant components, large construction machines, super express trains, etc. The weight of these heavy objects exceeds 30 ton, and therefore these ships are equipped with powerful cranes.

1.1.3 Other miscellaneous ships and marine transportation

Other miscellaneous types of ships include pilot boats, oil supply ships, water supply ships, military ships, submarines, fisherman ships, whale ships, fishery research ships, pleasure boats, motorboats, yachts, lifeboats, etc. Other ship classification methods include the shape and outer view of ships, supporting and propulsion type of ships, and length of ships (e.g., ultralarge ships: length $L > 300$ m, large ships: $L = 100-300$ m, and small ships $L < 100$ m). It is impossible to introduce everything here in detail because new types of ships are being constructed for a new purpose or for combined multiple purposes. In the preceding subsections, only representative examples are introduced.

The marine transportation efficiency of ships, as defined by the work rate (mass ton × velocity km/s) per unit engine horsepower (HP, 1 HP = 745.7 W), is very high (~300) although the moving velocity is usually low (< 50 km/h) [1]. Therefore, ships have very low CO_2 emission and are more economical than other transportation methods such as automobile, locomotive, and airplane. The emissions per unit power of automobiles, locomotives, and airplanes are approximately 10 times larger. However, the usage of a ship is limited to the low-speed region (< 50 km/h). Generally, a ship capable of achieving speeds greater than 30 km/h is called a high-speed ship. In 2014, the total mass × distance of cargos carried by ships worldwide was estimated to be more than 96,000 Gton·km, according to Shipping Review and Outlook, Spring 2016 [1]. However, the total mass × distance of cargos carried by airplanes worldwide in 2014 was estimated to be approximately 75 Gton·km [1]. Compared to airplanes, ships carry approximately 1000 times more cargo every year. For example, in Japan, 44% of domestic transportation is performed by ships, and 99% of cargos imported to or exported from Japan are carried by ships. Therefore, ships are very important transporters of cargos.

1.2 Marine diesel engines inside ships

There are numerous types of ship propulsion systems. Wind capture propulsion and paddle steamer propulsion are classical ones. Currently, most ships are driven by a screw propeller, which is illustrated in Fig. 1.2.1. The propeller wings are set with an angle of attack as shown in the figure. Therefore, lifting or propulsion force is induced owing to the water flow by the rotation of the screw propeller. Simultaneously, the drag force not shown in the figure is induced and transferred to the engine. When the ship begins to move, an axial drag force by axial water flow is induced. As a result, an axial driving force which equals to the lifting force minus the drag

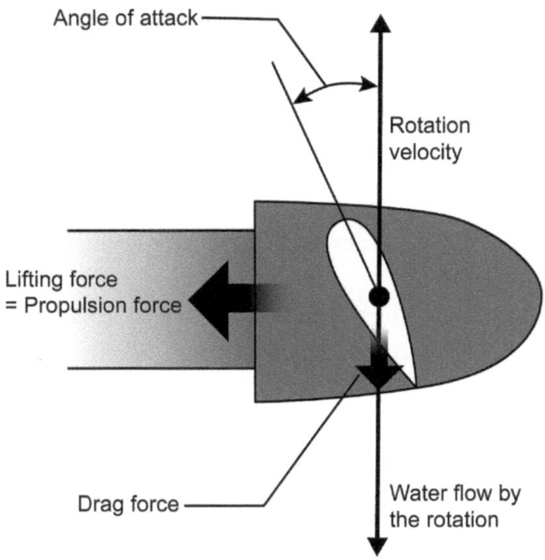

Figure 1.2.1 Schematics of screw propeller and acting forces [2].

force totally acts on the axis of the propeller. The axis is connected to the main engine inside the ship. Consequently, the ship moves.

After the invention of steam engines, the main engines of commercial ships have been steam engines similar to reciprocating engines of locomotives. The engine has a piston and cylinder; where the piston is moved by steam power linearly, and its kinetic energy is converted to rotational motion and utilized. This was the earlier type of commercially available reciprocating marine engine. However, the efficiency of this type of engine is not so high; hence, the steam turbine engine was introduced as the next commercialized marine engine. In this system, steam is directly injected into the rotating turbine blades. The system can generate rotational motion directly, and its compact size and large power generation are considered to be significant advantages. Next, marine diesel engines became commercialized. After the 1980s, steam turbines were used only for special purposes, such as in LNG ships, because of their higher fuel consumption and lower thermal efficiency compared with marine diesel engines.

Today, most ships use diesel engines, except for small boats that are driven by gasoline engines. The typical structure of diesel engine is shown

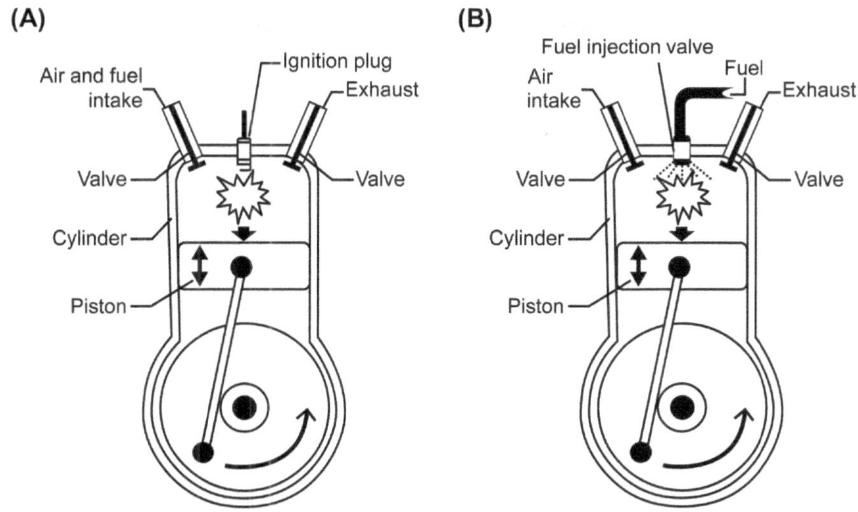

Figure 1.2.2 Basic structure of internal engines. (A) Gasoline engine and (B) diesel engine [2].

in Fig. 1.2.2 and compared with typical gasoline engines [2]. In the figure, an only single cylinder of each engine is schematically shown. In the gasoline engine as shown in Fig. 1.2.2A, a mixture of gasoline and air is introduced to the cylinder, and combustion occurs by igniting that mixture with a spark plug. The mixture of air and fuel is formed by fuel injection at the air intake or a carburetor. Recently, we have direct-injection gasoline engines. In such engines, gasoline is directly injected into the cylinder. In the diesel engine as shown in Fig. 1.2.2B, the air is introduced to the cylinder followed by fuel injection into the cylinder, and combustion occurs by compressing the volume of the cylinder; spark plugs are not used. Diesel engines have a higher pressure ratio, larger power output, and simpler structure, and they are tougher than gasoline engines. A diesel engine can also use different kinds of fuel oils: GO, marine diesel oil (MDO), intermediate fuel oil (IFO), marine fuel oil (MFO), and heavy fuel oil (HFO) are used for marine diesel engines. According to the Japanese Industrial Standards of fuel, GO is light oil, MDO is A heavy oil (sulfur $< 0.5\%$ or 2.0%), IFO is B heavy oil (sulfur $< 3.0\%$), and MFO and HFO are C heavy oil (sulfur $< 3.5\%$).

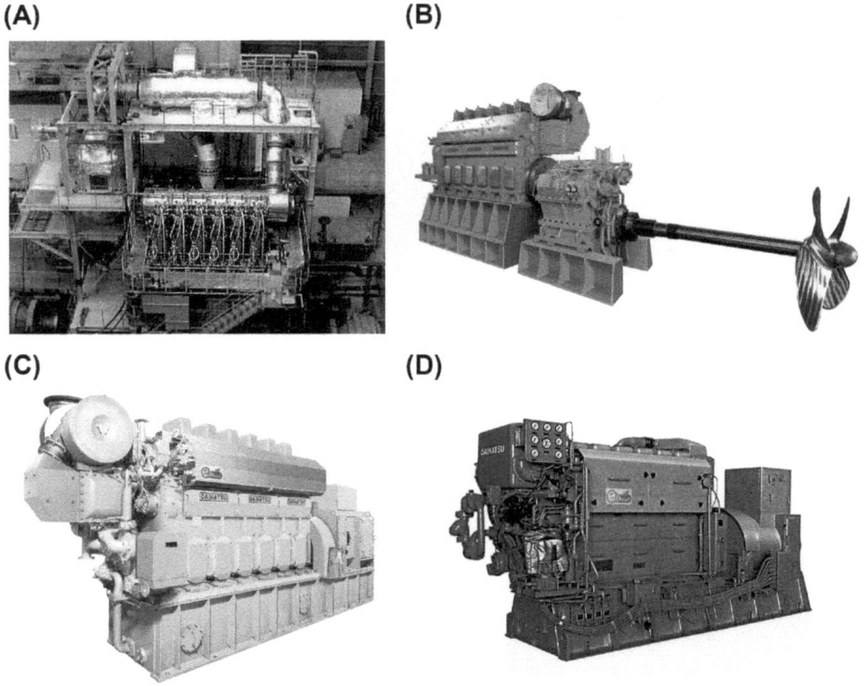

Figure 1.2.3 Typical marine diesel engines [5,6]. (A) Hitachi-MAN B&W, 6S46MC-C7, (B) Daihatsu Diesel MFG Co., Ltd., 6DKM-26e, (C) Daihatsu Diesel MFG Co., Ltd., 6DE-23, and (D) Daihatsu Diesel MFG Co., Ltd., 6DL-16Ae.

These abbreviations, GO, MDO, MFO, and HFO, are repeatedly used throughout this book. Furthermore, low-quality heavy oil, such as bunker oil, can also be used as marine engine fuel. The fuel names of low sulfur fuel oil (LSFO) of sulfur $< 0.1\%$ and high sulfur fuel oil (HSFO) of sulfur $> 3.5\%$ are also used.

Fig. 1.2.3 and Table 1.2.1 show photographs and typical specifications of marine diesel engines, respectively. There are low-speed, medium-speed, and high-speed engines. The rotation rate is usually constant and at a lower value than automobile rotation rates. Two-cycle and four-cycle engines exist. The main propulsion engines and power generators or auxiliary engines are installed in ships, and MFO or HFO are often used as fuels because of their lower cost. The mechanism of diesel engines is explained in the next section.

Table 1.2.1 Typical specifications of marine diesel engines [5,6].

Classification	Low speed < 300 rpm	Medium speed 300–1000 rpm		High speed > 1000 rpm
Manufacturer and type	Hitachi-MAN B&W 6S46MC-C7	Daihatsu Diesel MFG. 6DKM-26e	Daihatsu Diesel MFG. 6DE-23	Daihatsu Diesel MFG. 6DL-16Ae
Purpose of the engine	Main propulsion	Main propulsion	Power generator	Power generator
Number of cylinder	6	6	6	6
Rotation speed, rpm (max) cycle	107 (111) 2	750 4	720 (900) 4	900 (1200) 4
Cylinder diameter (bore), mm	460	260	230	165
Stroke length, mm	1932	380	320	210
Stroke:bore ratio	4.2:1	1.46:1	1.39:1	1.27:1
Power output, kW (max) [Generator power]	6100 (6780)	1810	1280 (1500) [1215 (1425)]	350 (530) [320 (480)]
Photograph	Fig. 1.2.3A	Fig. 1.2.3B	Fig. 1.2.3C	Fig. 1.2.3D

1.3 Mechanism of diesel engines

1.3.1 Heat engines

Mechanism of combustion engines are explained [4,7]. A heat engine is a machine that converts heat energy into mechanical energy to generate power. This type of engine is classified as either an external combustion engine or an internal combustion engine. An example of an external combustion engine is a steam turbine engine. A boiler is located away from the engine, fuel is burned in burners inside the boiler, and the heat converts water into steam. Steam is introduced into the turbine to rotate an electric power generator. An example of an internal combustion engine is a diesel engine in that it burns a mixture of fuel and air directly in cylinders and then uses the expansion of the combustion gas to move the piston to generate electric power. This book mainly focuses on emission control for internal combustion diesel engines in ships or marine diesel engines.

Internal combustion engines have several advantages over external combustion engines. Because the fuel burns directly in the cylinder of the internal combustion engine, the heat loss is smaller, the thermal efficiency is higher, and fuel consumption is lower. Also, because there is nothing in the system that must be installed separately, such as a boiler, the weight and volume of the entire engine system are relatively small. The total number of engine parts can be reduced. Furthermore, it can be started up in a short period of time and is economical because fuel is not wasted before and after startup; thus, the cruising distance of ships increases. However, internal combustion engines also have several disadvantages. Because the working pressure achieved by combustion is more likely to induce many shocks and vibrations in the engine, each part must be constructed with high-quality materials. Therefore, a large flywheel is usually required to smooth the rotation of the engine. In addition, because many parts are exposed to high humidity and high-pressure conditions, special attention should be paid to lubrication.

The diesel engine was invented by a German engineer named Dr. Rudolf Diesel approximately 125 years ago. During the early stages of development, to burn fuel, the air is compressed to a high pressure (~10 MPa) and then heated to a high temperature (500–550°C), causing the atomized fuel injected into the combustion chamber or cylinder to self-ignite spontaneously. In more current technology, on injecting fuel from a fine nozzle (diameter 0.3–0.8 mm) by applying high pressure of 30–200 MPa, oil particles are atomized at a high speed and burned immediately. Fuel atomization is the breakup of the fuel, i.e., fuel changes into a lot of very small particles when it is injected into the cylinder. Proper atomization facilitates the starting of the burning and ensures that each minute particle of the fuel is surrounded by oxygen particles that can ignite the fuel.

The advantages of the diesel engine are as follows. It is not necessary to perform electric ignition, as it is usual with the gasoline engine. It is only necessary for the diesel engine to compress air, and therefore the compression pressure can be increased. Because the thermal efficiency of the internal combustion engine increases as the compression pressure increases, the diesel engine has the highest thermal efficiency among all other heat engines. An ignition device is unnecessary, and if the valve adjustment is accurate, preignition will not occur. Because both the compression pressure and temperature are high, low-grade heavy oil can be used for ships.

Furthermore, because fuel and lubricant oil consumption is smaller, the operating cost is lower. However, the diesel engine does have several disadvantages. Because the burning pressure is high, only high-quality materials should be used, and the parts must be well manufactured. As a result, the total weight increases. Because the fuel must be injected at a high pressure, the fuel injector becomes complex and difficult to handle.

1.3.2 Classification of diesel engines

From the viewpoint of usage, diesel engines are classified into industrial engines, farm engines, car engines, land engines, and marine engines. The industrial engine is often a lightweight type used for general purposes, such as fire pumps and lightning during load repair. The farm engine is for agriculture or civil engineering. The car engine is for automobiles, locomotives, etc. The land engine is a power generator for factories. A relatively large power engine (~500 kW) is currently used as an emergency power supply for various buildings, such as hospitals, office buildings, and nuclear power stations. The marine diesel engine is used for the main propulsion of ships (prime mover) or for onboard power generation. Leading manufacturers of marine diesel engines are MAN Energy Solutions in Germany, Wärtsilä Corporation in Finland, Mitsubishi Heavy Industries, Ltd. in Japan, Yanmar Co., Ltd. in Japan, Daihatsu Diesel MFG Co., Ltd. in Japan, and Caterpillar Inc. in the USA. Other manufacturers and engineering companies are Mitsui E&S Shipbuilding Co., Ltd. in Japan, Hitachi Zosen Corporation in Japan, Niigata Power Systems Co., Ltd. in Japan, Rolls-Royce Holdings plc in the UK, etc.

From the viewpoint of cylinder arrangement, a diesel engine can be classified as horizontal engine, vertical engine, V-type engine, W-type engine, radial engine, or inverted engine. The cylinders are placed horizontally in a horizontal engine and vertically in a vertical engine. In a V-type engine, cylinders are placed diagonally; in a W-type engine, cylinders are placed in three W-shaped rows. In a radial engine, cylinders are arranged in a star shape. The inverted engine has inverted cylinders. The horizontal engine is used in land and vehicles, whereas the vertical engine is widely used in land and marine applications. The V-type engine is used for land, marine, and vehicles when greater power is required in a narrow space. The W-type engine, radial engine, and inverted engine are used in airplanes.

From the viewpoint of the cooling method for cylinders, diesel engines can be classified as air-cooled engines, liquid-cooled engines, and water-cooled engines. In an air-cooled engine, the cylinders are cooled by applying airflow. In a liquid-cooled engine, the cylinders are cooled by circulating liquid, such as oil. In a water-cooled engine, cylinders are cooled by freshwater or seawater. Air-cooled engines are widely used in airplanes and general-purpose machines and agriculture. Land and marine diesel engines utilize water for cooling. In marine engines, fresh seawater is used for direct cooling.

From the viewpoint of cyclical motion, a diesel engine can be classified as either a four-cycle engine or a two-cycle engine. For combustion to occur inside the cylinders, there are engines that burn every two revolutions of the crankshaft and those that burn once in every revolution. Because one revolution consists of two strokes of the piston, the former is called four-cycle engine or four-stroke engine, whereas the latter is called two-cycle engine or two-stroke engine.

Furthermore, from the viewpoint of engine rotation speed, diesel engines can be classified into low-speed engines, medium-speed engines, and high-speed engines. Typical examples of the specifications of marine diesel engines are already shown in Table 1.2.1. In large engines containing cylinders with large inner diameters, the rotational speed is relatively low. Conversely, small engines containing cylinders with small inner diameters have higher rotational speeds. Although the ratio of the stroke to the cylinder diameter (bore), i.e., the stroke/bore ratio, varies and depends on the application of the engine, it is roughly 1—4. The two-cycle engine is often used for low-speed applications, and the four-cycle engine is used for medium- and high-speed applications.

It is difficult to determine a strict range of marine diesel engine rotational speed; therefore, a general idea is presented here. An engine with a maximum rotational speed of 300 rpm is usually considered to be a low-speed engine. Most large two-cycle, low-speed engines are recognized as economical and reliable prime movers for large merchant and cargo ships. The cylinder diameter for these types of engines is typically 0.3—1.0 m, the stroke is 1—3.5 m, the stroke/bore ratio is 2.5—4.2, and the engine is fueled by MFO or HFO. Engines with a maximum rotational speed in the range of 300—1000 rpm are classified as medium-speed engines. They are four-cycle engines used for ferries, RO/RO ships, passenger ships, and medium-size cargo ships. Their cylinder diameters are typically 0.26—0.6 m, their stroke is 0.3—0.6 m, the stroke/bore ratio is 1—1.5, the fuel consumption

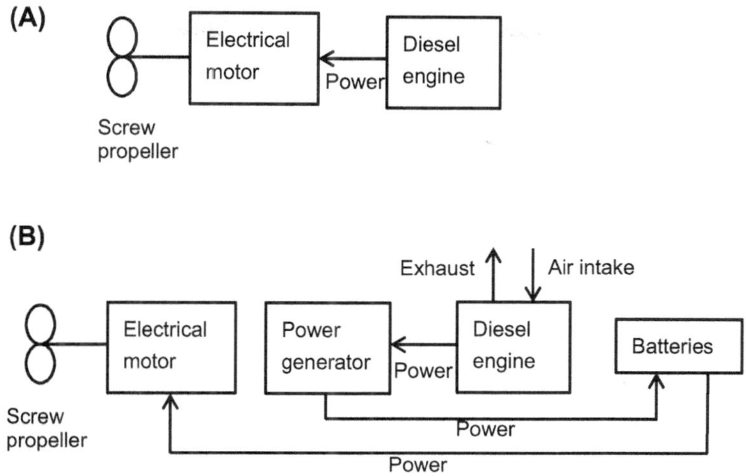

Figure 1.3.1 Schematic diagram of diesel engine driven electrical motor propulsion systems. (A) Direct electrical propulsion system and (B) electrical propulsion system using electrical energy stored in batteries.

rate is ~200 g/kWh, and they are fueled by GO, MDO, MFO, or HFO. Engines with a maximum rotational speed greater than 1000 rpm are classified as high-speed engines. These engines are used for high-speed ships and special operation ships. Their cylinder diameters are typically 0.15—0.25 m, the stroke is 0.15—0.21 m, the stroke/bore ratio is 1—1.4, and they are fueled by GO or MDO.

High-speed engines are sometimes used to drive an electric generator on the ship. Fig. 1.3.1 shows the schematic of the systems. Fig. 1.3.1A shows a direct-drive system, and Fig. 1.3.1B shows an indirect-drive system with batteries. In the system shown in Fig. 1.3.1B, the induced electric power is temporarily stored in batteries inside the ship, and the electric motor (prime mover) is driven by the power. This type of electrical propulsion system has been used in submarines for a long time and has recently begun to be utilized in ordinary ships.

1.3.3 Four-cycle and two-cycle engines

It has already been mentioned that there are four-cycle and two-cycle diesel engines. The motion mechanism of these engine types is explained below.

The movement of a four-cycle diesel engine is shown in Fig. 1.3.2A. In the engine, combustion occurs once in every two rotations of the wheel. The indicator diagram of the four-cycle engine in Fig. 1.3.2B shows the

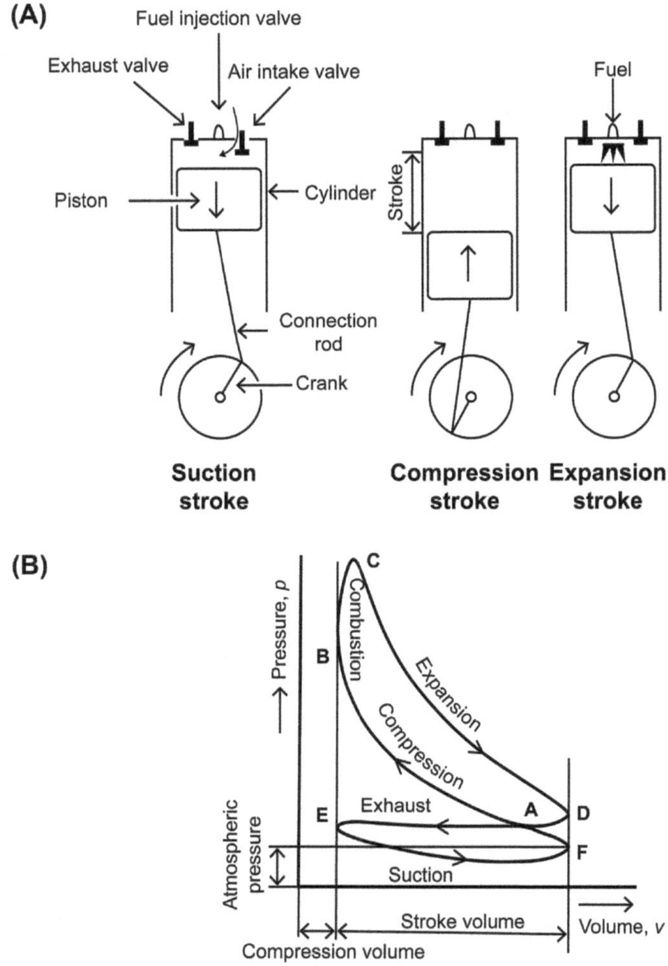

Figure 1.3.2 Movement and indicator diagram of four-cycle diesel engine. (A) Movement and (B) indicator diagram.

relationship between the volume v and pressure p in the cylinder. The processes are explained as follows:

1. Suction stroke (first stroke)

In this process, known as the suction stroke, the air is sucked into the cylinder. The exhaust valve is closed and the air intake valve is opened to introduce air into the cylinder, either through natural suction or the turbocharger. The indicator diagram shows the pressure change inside the cylinder, which is lower than the atmospheric pressure line as indicated by the solid curve (E—F).

2. Compression stroke (second stroke)

In this process, the intake valve and exhaust valve are closed, and the piston compresses the air upward. This process is called the compression stroke. The indicator diagram shows the gradual increase in pressure for compression as shown in the curve (F—B).

3. Expansion stroke (third stroke)

When the piston reaches the upper end of the cylinder, fuel is injected and the air/fuel mixture is ignited by the compression action at the same time. Then, the mixture burns and pushes down the piston. This process is called the expansion stroke. B—C in the indicator diagram shows the rapid change in combustion pressure, and C—D shows the gradual decrease in pressure.

4. Exhaust stroke (fourth stroke)

In this process, before the piston reaches the lower end of the cylinder, the exhaust valve is opened at position D in the indicator diagram and the exhaust gas is discharged. When the piston reaches the upper end at state E of the diagram, it closes. In the indicator diagram, D—E is higher than the atmospheric pressure line. This process is called the exhaust stroke.

The movement of a two-cycle diesel engine is shown in Fig. 1.3.3A. In this engine, combustion occurs once in every single rotation of the wheel. The indicator diagram of the two-cycle engine is shown in Fig. 1.3.3B. The diagram is explained as follows:

1. Compression and combustion (first process)

When the piston gradually rises in the cylinder from the lower end, it aspirates air from the aspiration valve to the crank chamber. Also, after the piston closes the sweeping port and the exhaust valves are closed, it compresses the air, and combustion begins after the fuel is ignited at the end of its stroke.

2. Exhaust and suction stroke (second process)

The combustion lowers the piston. During this process, when the upper end of the piston comes to the sweeping port, the compressed air in the sealed crank chamber ejects into the cylinder through this port and discharges the exhaust from the exhaust valve.

The advantages of the four-cycle diesel engine are as follows. Fuel loss is low and thermal efficiency is high. Also, the cleaning action inside the cylinder is reliable, and the life of the cylinder is long. However, the four-cycle engine does have several disadvantages. Because of its complex structure and uneven rotation, a large flywheel is required. Also, if it has the same volume as a two-cycle engine, its power output will be smaller.

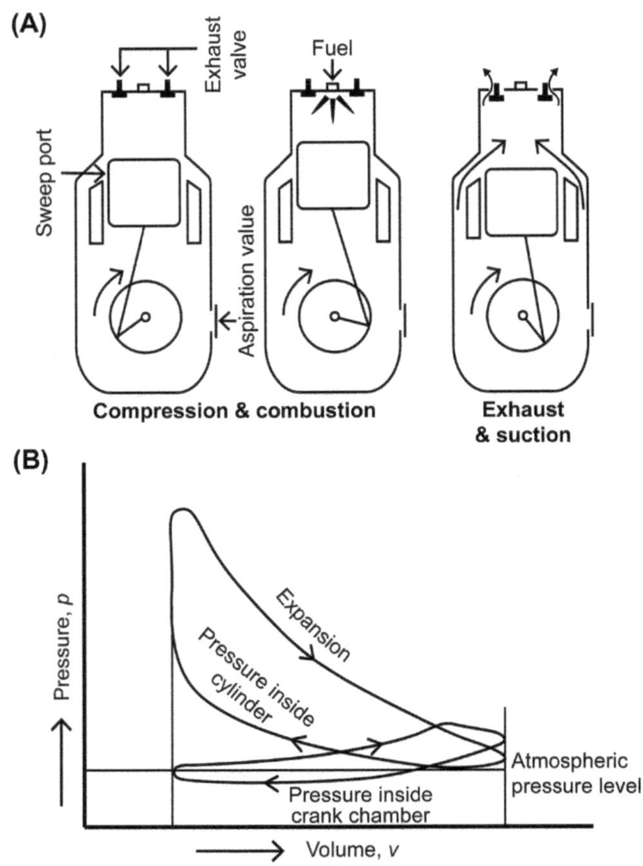

Figure 1.3.3 Movement of two-cycle diesel engine. (A) Movement and (B) indicator diagram.

The advantages of the two-cycle engine are as follows. High output can be obtained with the same cylinder volume. Compared with the four-cycle engine that burns once in every two revolutions, the two-cycle engine burns only once per revolution, and it theoretically should be able to provide twice the output with the same cylinder volume; in reality, it is more than 1.5 times. The design is easy to construct, is easy to handle, and features smooth rotation, and the flywheel can be small. However, the design of the two-cycle engine has several disadvantages as well. The exhaust tends to become insufficient. A device that sends high scavenging pressure, such as blower, is necessary in large engines. Because more than 20% of the stroke is used for the exhaust port, the stroke becomes less effective. Two-cycle engines require more costly parts than four-cycle engines.

Exploiting the advantages of both designs, the four-cycle engine is often adopted for medium-/small-size marine diesel engines and high-speed engines used for power generation, and the two-cycle type is adopted for large marine diesel engines for ship's propulsion. It is noted that the rotation of two-stroke engine is generally reversible and maneuvering is easy, although the rotation of two-cycle or two-stroke engines is nonreversible and maneuvering is done by other means.

1.3.4 Common rail fuel injection technology

In a diesel engine, from the perspective of environmental protection, fuel injection is essentially a key technology for not only the engine operation but also the fuel consumption and emissions because it is related with combustion improvement. Fuel injection technologies can be classified into mechanical and electronic control technologies.

One of the mechanical control technologies is the variable valve timing mechanism. Through this technology, the combustion temperature can be lowered when the intake valve is closed before reaching the bottom dead center to reduce the effective compression ratio. As a result, NO_x emission is reduced. However, merely closing the intake valve causes a lack of intake air for combustion, and output power becomes insufficient. To resolve this problem, a high-pressure supercharger is required.

To optimize the fuel injection timing, electronic control technologies are used recently for diesel engines. The electronic control technologies can be classified into unit fuel injection and common rail fuel injection technologies. The unit fuel injection system is a high-pressure integrated direct fuel injection one for diesel engines, combining the injector nozzle and the injection pump in a single component. In the unit injection system, a fuel is supplied to a solenoid valve and an injection nozzle by an injection pump. The unit injection system has a similar structure to that of a traditional injection system. The solenoid valve is installed, having a distance from an injection nozzle. As its advantages, a traditional injector and a cylinder head can be used in the unit injection system, the structure of which is simple. However, as a disadvantage, injection pressure depends on the rotational speed of the engine. Moreover, it is difficult to achieve multiple injections because the response is not rapid owing to the long distance between the solenoid valve and the injection nozzle. On the other hand, Fig. 1.3.4 shows a schematic of a common rail fuel injection system along with a photograph of its installation in a marine diesel engine [8]. Because a solenoid valve exists inside each injector, the distance between each valve and

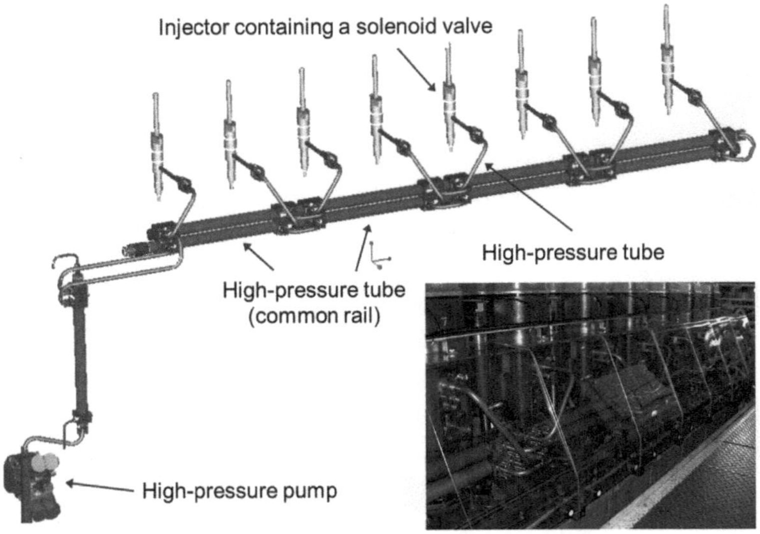

Figure 1.3.4 Schematic of common rail fuel injection system along with a photograph of its installation in a marine diesel engine [8].

injection nozzle is shorter. As a result, the response for the injection is so rapid. As another advantage of a common rail fuel injection system, the fuel injection can take place under arbitrary timing and high pressure because the high pressure can be accumulated in a common rail high-pressure tube. High-pressure injection gives environmental, power, and fuel consumption benefits over earlier lower pressure fuel injection, by injecting fuel as a larger number of smaller droplets, having a much higher ratio of surface area to volume. This provides improved vaporization from the surface of the fuel droplets, and so more efficient combining of atmospheric oxygen with vaporized fuel delivering more complete and cleaner combustion with lower NO_x and PM emissions. Further, solenoid or piezoelectric valves make possible fine electronic control over the fuel injection timing and quantity, and the higher pressure that the common rail technology makes available provides better fuel atomization. In the common rail system, because arbitrary timing and multiple injections such as pre- and post-injections are possible, the injections take place several times in a cycle to reduce NO_x emission. In more detail, a pilot injection that is preinjection with a small amount of fuel before the main injection takes place. NO_x emission is reduced because the combustion by the pilot injection can reduce ignition delay and can prevent rapid increases of temperature and pressure. In addition, an after injection that takes place with a small amount of fuel followed by the main injection is also used in the common rail system. PM and smoke can be treated by the combustion by the after injection that stirs gas inside of a cylinder and PM reduction by fine fuel injection due to the high pressure. The test results reveal that smoke in marine diesel emissions can be reduced through a high-pressure common rail injection [8]. No smoke emissions are observed at an injection pressure of 160 MPa. Further optimization of the timing and pressure of the injection for every engine rotating speed and load can greatly reduce both NO_x and PM emissions. As a disadvantage of the system, the structure is slightly complicated, and a traditional injector and cylinder head cannot be used without redesigning.

1.4 Targets of emission control

In this book, practical and fundamental aspects of various types of emission control technologies are treated. The main targeted emissions from marine diesel engines are nitrogen oxide (NO_x), sulfur oxide (SO_x), PM, volatile organic compound, and CO_2. This chapter provides an introduction to the different types of ships and the fundamentals of marine diesel engines, their mechanism, and emissions. In the subsequent chapters, not only the principles and design considerations of emission control technologies for marine and automobile applications but also the fundamentals of environmental engineering are treated. Laboratory, pilot-scale, and full-scale operation examples of aftertreatment systems are explained. The achievements of our plasma superclean diesel engine project are also described. Furthermore, future prospects for marine diesel engine emission control are explained. $PM_{2.5}$ emission and health effect; CO_2 control; past, current, and future market sizes of aftertreatment systems; and the operation and maintenance of various aftertreatment systems are treated as important topics. As described above, the emergency targets to be cleaned, which are emitted by marine diesel engines, are mainly NO_x, SO_x, and especially PM. Fig. 1.4.1 shows numerically predicted geospatial distribution of $PM_{2.5}$ pollution due to oceangoing ships in recent years [9]. In the figure, the unit of $PM_{2.5}$ concentration is shown in $\mu g/m^3$ and values represent concentration due to shipping compared to an environment without shipping emissions. $PM_{2.5}$ is very harmful to humans, and its health effects are treated in Chapter 5.

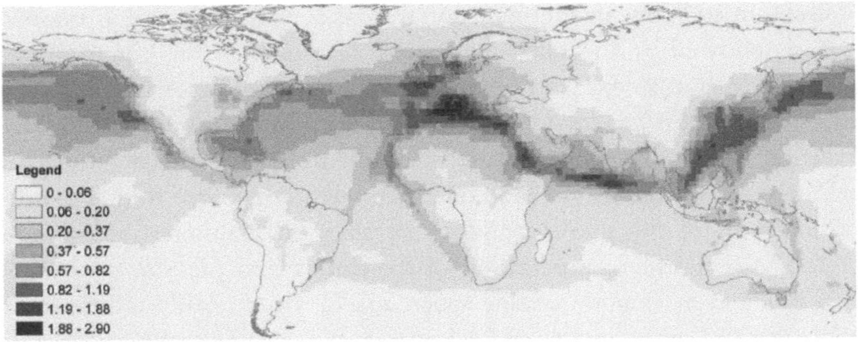

Figure 1.4.1 Numerically predicted geospatial distribution of $PM_{2.5}$ concentration (unit is $\mu g/m^3$) caused by pollution due to oceangoing ships in the recent years [9].

References

[1] T. Kawasaki, Zukai Yoku Wakaru Saisin Senpaku No Kihon to Shikumi (Illustrated Basics of the Latest Ships and Mechanism), 3rd ed., Shuwa System Co. Publ., 2014 (in Japanese).
[2] T. Kawasaki, Koredake Senpaku (Book on Marine Engineering), 2nd ed., Shuwa System Co. Publ., 2014 (in Japanese).
[3] O.S.K. Mitsui, Lines Pamphlet, Iroiro na Fune (Various Types of Ships), 2018 (in Japanese), http://www.mol.co.jp/iroiro_fune/pdf/iroiro2018.pdf.
[4] A. Ayub, Marine Diesel Engine, CRC press, 2010.
[5] M. Murakami, H. Tanaka, M. Fukuyama, T. Fujibayashi, S. Okazaki, J. Shibata, H. Tanaka, S. Hiraiwa, R. Kobayashi, Y. Ezu, Development of marine SCR system (separate type) for large two-stroke diesel engines, Hitz Tech. Rev. 75 (1) (2014) 2–9 (in Japanese).
[6] Daihatsu Diesel MFG Co., Ltd. website, 2018, http://www.dhtd.co.jp/en/products/catalog/.
[7] S. Hasegawa, Hakuyo Diesel Kikan Kyohan (Textbook of Marine Diesel Engines), Seizando Syoten, 2003 (in Japanese).
[8] K. Hanamoto, Daihatsu Diesel's energy saving technologies, J. Japan Inst. Mar. Eng. 49 (2) (2014) 207–211 (in Japanese).
[9] J.J. Winebrake, J.J. Corbett, E.H. Green, A. Lauer, V. Eyring, Mitigating the health impacts of pollution from oceangoing shipping: an assessment of low-sulfur fuel mandates, Environ. Sci. Technol. 43 (13) (2009) 4776–4782.

CHAPTER 2

Emission regulations

Contents

2.1 Establishment of MARPOL 73/78 Convention

The voyages of ships are a global issue. Ships from various countries arrive and dock at ports or terminals. If the standards on size, facilities, safety, and environmental impact of ships vary from country to country, port reception services/facilities become confusing, which may lead to accidents and environmental pollution. Therefore, worldwide standards and regulations for all ships are necessary for maritime safety and environmental protection. The International Maritime Organization (IMO) discusses and regulates the conventions on maritime safety, efficient voyage, and prevention of marine pollution. The need for an international convention that covers the prevention of pollution by ships from operational, accidental, or emissions causes has resulted in the International Convention for the Prevention of

New Technologies for Emission Control in Marine Diesel Engines
ISBN 978-0-12-812307-2
https://doi.org/10.1016/B978-0-12-812307-2.00002-X

Pollution from Ships, also known as MARPOL, which is an abbreviation for marine pollution.

In general, the MARPOL convention is referred to as MARPOL 73/78, which means MARPOL 1973 as amended by the protocol of 1978. From a historical point of view, in an early stage of marine environmental protection, the International Convention for the Prevention of Pollution of the Sea by Oil was an international treaty signed in London on May 1954, known as OILPOL 54. This convention aimed to control oil bilge from operating tankers to prevent ocean pollution. However, in 1967, the tanker named Torrey Canyon ran aground in the English Channel and released 120,000 ton of crude oil into the sea, making it one of the largest oil spills in the world. This incident and growth in the size of tankers raised awareness of environmental protection in the maritime field. Although OILPOL 54 was applied only to new tankers, new international convention for all tankers was required. In line with this, the OILPOL 54 convention was amended repeatedly in 1962, 1969, and 1971 by the Inter-Governmental Maritime Consultative Organization, the predecessor of IMO. In 1973, MARPOL 73 convention was adopted. The convention targets not only heavy oil but also all kinds of oils, chemicals, and harmful substances to prevent marine pollution. However, the convention was not implemented because of unresolved issues on differences in technical levels among the countries. In 1976—77, a series of tanker accidents occurred. Thus, more stringent action was demanded to avoid accidental and operational oil pollution. In 1978, an amended convention known as MARPOL 73/78 was adopted with technical annexes, Annex I—V. In 1997, a protocol was adopted to amend the convention, and Annex VI was newly added. The new convention includes regulations aimed at preventing and minimizing pollution from ships, which includes both accidental pollution and that from routine operations. The regulations were issued against nitrogen oxide (NO_x), particulate matter (PM), carbon monoxide (CO), and nonmethane hydrocarbons (NMHCs) emissions. The convention includes six technical annexes, Annex I—VI, which are described in A.2 of Appendix.

2.2 NO_x regulations

2.2.1 Pollution of NO_x and photochemical smog

Although in general NO_x includes nitric oxide (NO), nitrogen dioxide (NO_2), nitrous oxide (N_2O), dinitrogen pentoxide (N_2O_5), etc., it mainly refers to NO and NO_2. In combustion processes, NO_x formation is inevitable because thermal NO_x is formed by the oxidation of N_2 by O_2 in the air. NO_x is an unwelcome byproduct by combustion because it is recognized as air pollutant and induces secondary pollutants such as photochemical smog. From a historical point of view, fossil fuel combustion emissions that include air pollutants have been a problem since the Industrial Revolution. During this period, environmental problems were caused by air pollution in the form of smog. People were already aware of the problem early in the 1800s. For instance, in Sheffield, UK, open-air burning of coke is prohibited. The first photochemical smog emerged in Los Angeles, USA, in the 1940s. The popularity of automobiles and the basin-shaped geography of Los Angeles caused the photochemical smog. As an example of serious historical incidents, in December 1952, people experienced the most serious smog disaster in London. This is known as Great Smog of London or Great Smog of 1952, in which more than 12,000 London citizens are considered to die by exposure. It is said that the main cause of the disaster was SO_x released by burning fossil fuel such as coal; NO_x was emitted much as well. At that time, the cold weather forced London residents to use and burn more coal than usual for heating. In addition, numerous coal-fired power stations and the shifting of transportation from electrical trams to diesel autobuses also contributed to the disaster. Since then, people began paying attention to environmental issues, and environmental legislation and regulations have been imposed since 1952, such as the City of London (Various Powers) Act 1954 and the Clean Air Act 1956 and 1968 in the United Kingdom. In the United States, the Air Pollution Control Act of 1955, the Clean Air Act of 1963 and 1970, and the Clean Air Act Amendment of 1977 and 1990 are specific milestones. In particular, the Clean Air Act of 1970, known as the Muskie Act, was a very stringent regulation at the time. Since then, many kinds of regulations have been enacted across the world to control air pollutants. In Japan, the first evidence of photochemical smog emerged in July 1970 in Tokyo, and then regulations have been enacted.

Photochemical smog means an atmospheric state where a mixture of gaseous photochemical oxidants, solid nitrate, and sulfate stays at high concentrations and visibility becomes poor. NO_x is a cause of photochemical oxidants and acid rain. Fig. 2.2.1 shows a diagram of photochemical oxidants formation. The mechanism of photochemical oxidants formation is described as follows. Firstly, NO emitted by combustion processes of automobiles, marine vessels, plants, etc. is oxidized to NO_2 in the atmosphere. Secondly, NO_2 is decomposed into NO and an O radical by the energy, $E = h\nu$, of ultraviolet (UV) light from the sun.

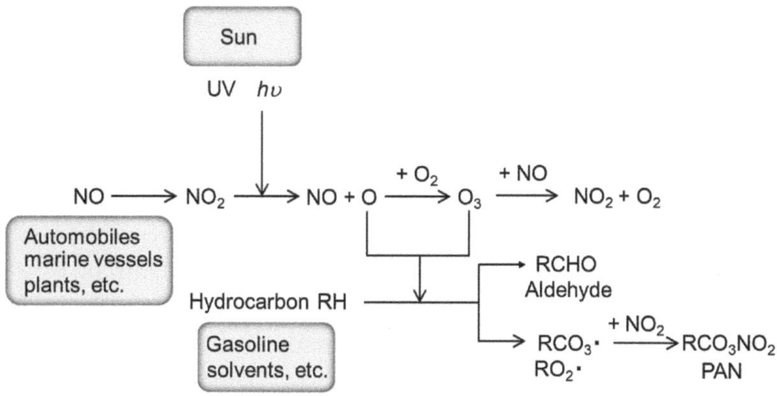

Figure 2.2.1 Diagram of photochemical oxidant formation.

$$NO_2 \rightarrow NO + O \tag{2.1}$$

The generated O radical reacts with O_2 in the air to form O_3 as follows:

$$O + O_2 \rightarrow O_3 \tag{2.2}$$

Part of the generated O_3 reacts with NO as

$$O_3 + NO \rightarrow NO_2 + O_2 \tag{2.3}$$

Finally, hydrocarbon (RH) or volatile organic compounds (VOC) from gasoline, the organic solvent, etc. react with the O radical and/or O_3 to form aldehyde and peroxides by chain reactions.

$$RH + O, O_3 \rightarrow RCHO,\ RO_2\cdot, RCO_3\cdot \tag{2.4}$$

where R is a nonmethane hydrocarbon group such as $-C_2H_5$, $-C_3H_7$, etc. because methane scarcely contributes to the photochemical reaction (2.4). Part of the peroxides reacts with NO_2 to form peroxyacetyl nitrate (PAN) (RCO_3NO_2) as follows:

$$RCO_3\cdot + NO_2 \rightarrow RCO_3NO_2 \tag{2.5}$$

In summary, NO_2 and VOC are incident materials or primary pollutants, and ozone (O_3), aldehyde (RCHO), and PAN (RCO_3NO_2) are the resultant materials or secondary pollutants called as photochemical oxidants. Therefore, NO_x and VOC concentrations in the atmosphere should be reduced to control the concentrations of result materials. Photochemical oxidants, especially aldehyde and PAN, are harmful to humans as they cause respiratory health problems including eye irritation and/or dyspnea.

Table 2.2.1 Ozone exposure concentrations and human health effects.

Concentration, ppm	Phenomena
0.01–0.02	One can smell a slight odor. Later, they get used to it.
0.1	One perceives an apparent odor that stimulates the nose and throat.
0.2–0.5	Visibility declines after 3–6 h of exposure.
0.5	One perceives a stimulation in the upper respiratory tract.
1–2	Headache, chest pain, thirst, and cough occur after a 2-h exposure, and chronic poisoning is caused by repeated exposure.
5–10	Heart rate increases, body pain, and anesthesia symptoms appear and continuing exposure leads to pulmonary edema.
15–20	Small animals die within 2 h.
50	Life-threatening for humans after 1 h.

Made based on Ref. [1].

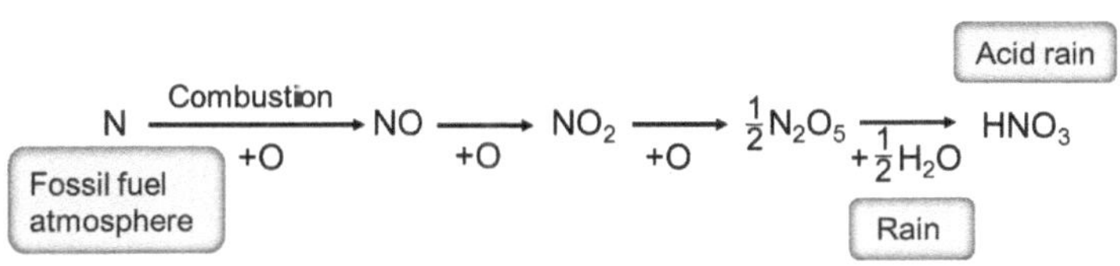

Figure 2.2.2 Diagram of acid rain formation due to NO_x.

Considered as the index for photochemical oxidants, O_3 at concentrations higher than 1 ppm poses threats to human health (Table 2.2.1). However, there are reports that low-concentration ozone (~0.1 ppm) has a sterilizing effect and gives a good influence to the human.

Next, formation of acid rain is explained. In general, rain with the potential of hydrogen pH ≤ 5.6 is called acid rain because pH becomes approximately 5.6 when CO_2 in the atmosphere is fully absorbed in the rain. Fig. 2.2.2 shows a diagram of acid rain formation due to NO_x. NO_2 oxidized from NO in the atmosphere reacts with oxygen in the air to form dinitrogen pentoxide (N_2O_5). This N_2O_5 and rain (H_2O) form nitric acid (HNO_3). This nitric acid solution is called acid rain due to NO_x. It is noted that acid rain is also caused by SO_x, which will be explained in Subsection 2.3.1.

2.2.2 NO_x regulations for automobile engines

Harmful NO_x emission has been a serious issue for diesel engines because NO_x formation is an inevitable consequence of the combustion process. Therefore, NO_x emission regulations have been imposed regionally and worldwide. In most cases, regulations were initially imposed on automobile emissions. These regulations have led to the development of emission control technologies, which are then applied to other forms of transportation or vehicle using diesel engines. Because they are related to marine engine emission controls, NO_x regulations for automobiles in Japan and Europe are reviewed in this section.

In Japan, automobile emission regulations for both gasoline and diesel engine cars have been imposed since 1966. These regulations have become more stringent year after year. For instance, the time-dependent regulations of emitted NO_x are shown in Fig. 2.2.3 in the case of diesel automobiles weighing more than 3.5 ton, such as buses and trucks. It is noted that the values in the horizontal axis represent the ratio of the mass of NO_x (unit is g) to output energy (unit is kWh) for a target engine. A regulation issued in 2005 is called New Long-Term Regulation and that issued in 2009 is called Post-New Long-Term Regulation, and a new regulation was issued further in 2010. The regulations were issued against NO_x, PM, CO, and NMHC emissions. It is noted that NMHCs are important as air pollutants among various VOCs because methane has little effect on the formation of photochemical oxidants. In the graph, the value decreases by 1.3 g/kWh from 2.0 g/kWh down to 0.7 g/kWh in 2009, 0.3 g/kWh from 0.7 g/kWh in 2016, and now comes the most stringent value of 0.4 g/kWh. In addition, the regional regulation was also imposed in 2017 on eight prefectures with a larger population of Tokyo, Osaka, Kanagawa, Aichi, Hyogo, Saitama, Chiba, and Mie.

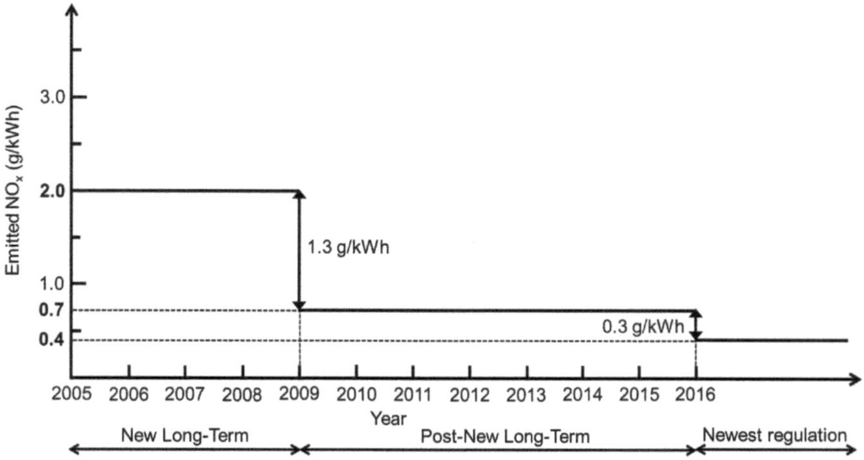

Figure 2.2.3 Recent NO_x regulation in Japan for diesel automobiles weighing more than 3.5 ton [2,3].

In Europe, the exhaust emissions of new automobiles in the European Union (EU) and European Economic Area (EEA) member states are controlled by European emission standards (Euro). For passenger cars, the regulations have been typically referred to as Euro 1, Euro 2, Euro 3, Euro 4, Euro 5, and Euro 6 since 1992. The objective of Euro emissions standards is to reduce the amount of harmful exhaust emissions such NO_x, CO, hydrocarbon (HC) compounds, and PM. Euro emissions standards have classifications for different automobile categories. As for approvals of new diesel passenger cars, Euro 1, 2, 3, 4, 5a, 5b, and 6 were issued in 1992, 1996, 2000, 2005, 2009, 2011, and 2014, respectively. Particle number (PN) regulation has also been required since the implementation of Euro 5b as explained later. As shown in Fig. 2.2.4, in recent regulations on NO_x for passenger diesel automobiles, the value is regulated from 0.25 g/km in Euro 4 down to 0.18 g/km in Euro 5a (0.07 g/km decrease) and down to 0.08 g/km in Euro 6 (0.07 g/km decrease). It is noted that the values are measured in g/km, that is, the ratio of the mass of emitted NO_x to mileage over the Worldwide harmonized Light vehicles Test Cycle (WLTC) using a chassis dynamometer procedure. The mass of emitted NO_x is measured as the equivalent one of NO_2. For heavy-duty trucks and buses, the regulations are typically referred to as Euro I, Euro II, Euro III, Euro IV, Euro V, and Euro VI. In recent regulations in Fig. 2.2.5 for heavy-duty trucks and buses with maximum weight over 3.5 ton, the value is regulated from

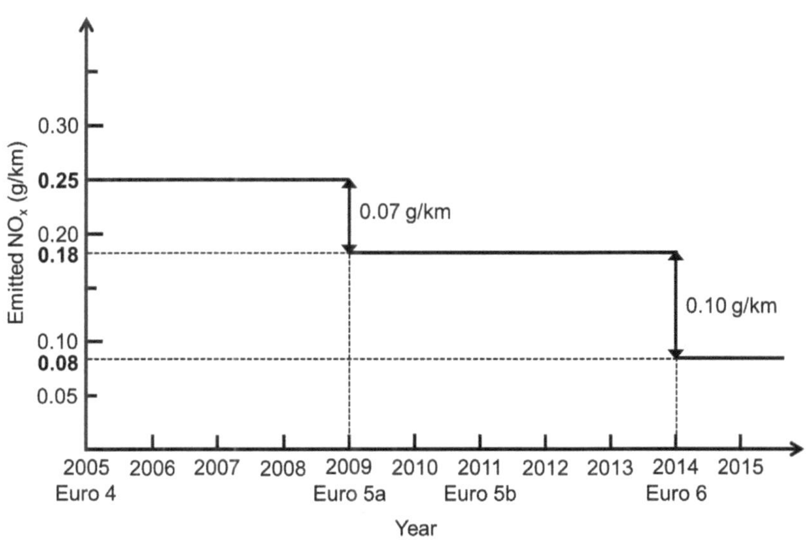

Figure 2.2.4 Recent NO_x regulation in Europe for passenger automobiles [4].

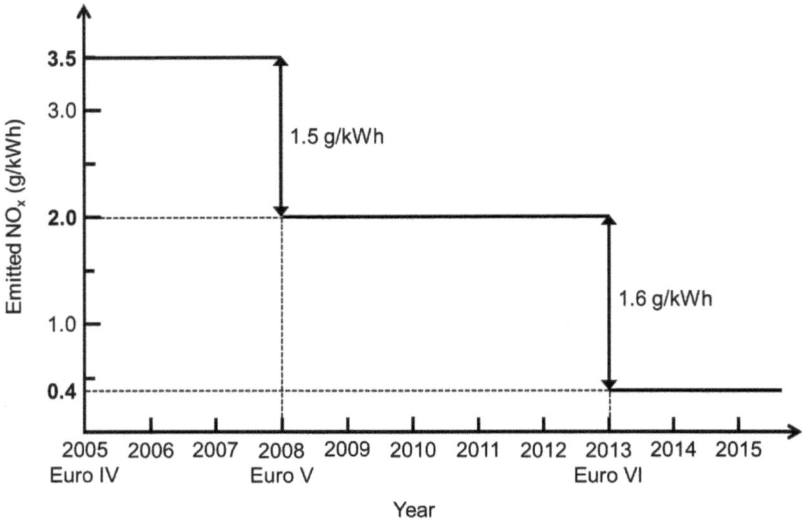

Figure 2.2.5 Recent NO_x regulation in Europe for heavy-duty diesel engines such as trucks and buses [5].

3.5 g/kWh in Euro IV in 2005 down to 2.0 g/kWh in Euro V in 2008, which corresponds to 1.5 g/kWh decrease, and down to 0.4 g/kWh in Euro VI in 2013, which corresponds to 1.6 g/kWh decrease. The values are measured in g/kWh, that is, the ratio of the mass of emitted NO_x to output energy of the engine. It is noted that emissions are usually tested over a test cycle using a chassis dynamometer procedure and expressed for passenger automobiles in g/km. However, this kind of test was difficult for heavy-duty trucks and buses because the body is too heavy to be placed on the inspection table of the chassis dynamometer; hence, the regulation has been expressed in g/kWh.

2.2.3 NO_x regulations for marine diesel engines

Recently, because the enforcement of NO_x regulations for automobiles is almost completed in western countries, the next target is directed toward marine diesel engines. In marine diesel engine emissions, NO_x concentration is usually high (500–1000 ppm), and SO_x (typically 200 ppm with MDO) is included because the engines are powered either by marine diesel oil (MDO) or marine fuel oil (MFO) containing sulfur.

The amounts of NO_x emissions from ships are regulated by MARPOL 73/78 convention Annex VI, Regulation 13. The regulations governing marine diesel engines are determined by the IMO, and NO_x emission

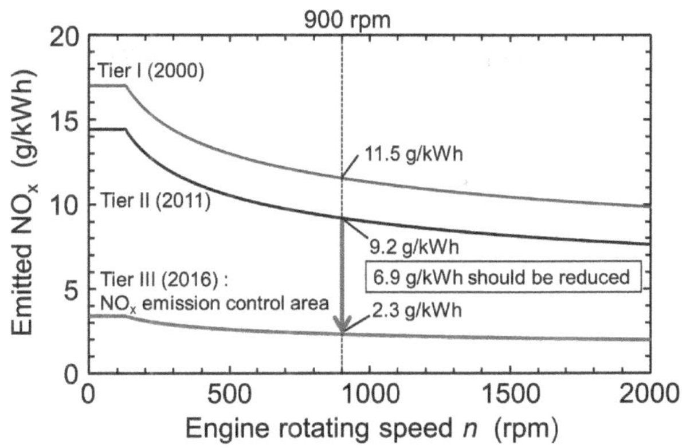

Figure 2.2.6 International Maritime Organization (IMO) emission standards and the requirement of NO_x reductions at Tiers I, II, and III [6].

standards have been imposed, as shown in Fig. 2.2.6. This regulation restricts the mass of NO_x (NO_2 equivalent mass) emission per engine output energy for a specified engine rotating speed n, as shown by the horizontal axis of the figure, at Tiers I in 2000, II in 2011, and III in 2016, which indicate the stages and enforcement year. In the curves in the figure, the values of the NO_x emission limits at Tiers I, II, and III are described as follows:

1. Tier I in 2000

   17.0 g/kWh at specified engine rotational speed $0 < n < 130$ rpm

   $45.0n^{-0.2}$ g/kWh at specified engine rotational speed 130 rpm $\leq n <$ 2000 rpm

   9.8 g/kWh at specified engine rotational speed $n \geq 2000$ rpm
2. Tier II in 2011

   14.4 g/kWh at specified engine rotational speed $0 < n < 130$ rpm

   $44.0n^{-0.23}$ g/kWh at specified engine rotational speed 130 rpm $\leq n <$ 2000 rpm

   7.7 g/kWh at specified engine rotational speed $n \geq 2000$ rpm
3. Tier III in 2016

   3.4 g/kWh at specified engine rotational speed $0 < n < 130$ rpm

   $9.0n^{-0.2}$ g/kWh at specified engine rotational speed 130 rpm $\leq n <$ 2000 rpm

   2.0 g/kWh at specified engine rotational speed $n \geq 2000$ rpm

Tier III is applied to ships equipped with diesel engines navigating in emission control areas (ECAs). The ECAs for NO_x regulation are determined as the North American area and the Caribbean Sea area. It is noted that the Baltic Sea and North Sea are included in the ECA for SO_x regulation as explained later, but not for NO_x regulation. Because large ships using diesel engines are operated internationally, there is a possibility of operating in ECA. Therefore, it is necessary for the diesel engines to comply with Tier III regulation. For instance, when a marine diesel engine operating at a rotational speed of 900 rpm is targeted, NO_x emission should be less than 11.5 g/kWh at Tier I, 9.2 g/kWh at Tier II, and 2.3 g/kWh at Tier III. Therefore, the required NO_x reduction from the regulation Tiers II to III is 6.9 g/kWh. This corresponds to a 75% reduction, which is a significant value. There is a possibility that marine diesel engines operated internationally that cannot satisfy this stringent Tier III requirement can no longer be manufactured after 2016. In light of the recent circumstances surrounding marine diesel emission as mentioned above, NO_x treatment is a very important issue. In 2011, most marine diesel engine companies succeeded in qualifying their engines to the Tier II NO_x emission standard. This was accomplished by means of improved combustion or engine adjustment. However, because Tier III is very stringent, effective aftertreatment technology was indispensable to achieve significant NO_x reduction. Urea-based selective catalytic reduction technology is a most promising solution and has been successfully applied to marine diesel engines, although it requires large urea solution tanks inside ships.

2.3 SO_x regulations

2.3.1 SO_x and acid rain

In general, SO_x includes sulfur monoxide (SO), sulfur dioxide (SO_2), and sulfur trioxide (SO_3). Because SO immediately oxidizes into SO_2 and cannot be decomposed at room temperature under atmospheric pressure. Therefore, it refers to SO_2 and SO_3. Although SO_2 is observed as gas or molecules, SO_3 is observed as PM. In combustion processes, sulfur in fuel leads to the formation of SO_x. SO_x formation is observed typically in cases involving fossil fuels such as petroleum and coal. SO_x is an unwelcome byproduct because it is considered an air pollutant along with NO_x. SO_x is the main cause of smog such as the Great Smog of London in 1952.

As well as NO_x, SO_x is also one of the causes of acid rain with $pH \leq 5.6$. Fig. 2.3.1 shows a diagram of acid rain formation due to SO_x. SO_2 in the atmosphere reacts with O_2 or O in the air to form SO_3. This SO_3 and rain (H_2O) form sulfuric acid (H_2SO_4) solution, which is acid rain

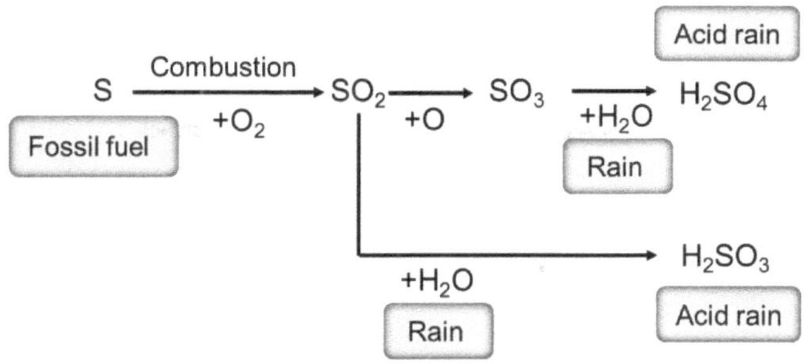

Figure 2.3.1 Diagram of acid rain formation due to SO_x.

due to SO_x. It is also possible that SO_2 and rain (H_2O) form sulfurous acid (H_2SO_3) solution, which is also acid rain due to SO_x.

Acid rain due to SO_x is an indirect contributor to diseases in human beings. However, SO_3 is a type of PM itself. SO_x in the atmosphere can cause respiratory diseases such as bronchitis and asthma. Thus, SO_x is directly harmful to human health and causes a serious issue that requires a solution.

2.3.2 Method to control SO_x emission

Similar to NO_x, harmful SO_x emission has been a serious problem for diesel engines because they use fossil fuels such as petroleum. Therefore, SO_x emission regulations have been imposed both regionally and worldwide. SO_x emissions can be reduced by lowering the sulfur content of fuel or by using low-sulfur fuels. In particular, diesel fuels for automobiles with substantially low sulfur content are called ultra-low-sulfur diesel (ULSD) fuel. The utilization of ULSD suppresses the possibility of catalyst poisoning by sulfur when catalysts are used for NO_x and PM treatments in emission control. Automobile diesel fuels in North America, Europe, and Japan have been replaced by ULSD since the mid-2000s. Therefore, in recent years, regulations directly addressing SO_x emissions have not been issued for automobiles because they are not needed.

2.3.3 SO_x regulation for marine diesel engines

For emissions from ships, the amount of SO_x emissions is regulated by MARPOL Annex VI, Regulation 14, which is explained in A.2 of Appendix. Instead of direct regulation of the amount of SO_x in emissions, the regulation sets a limit on the allowable sulfur content in fuels. The limits become more stringent in stages and differ according to the area,

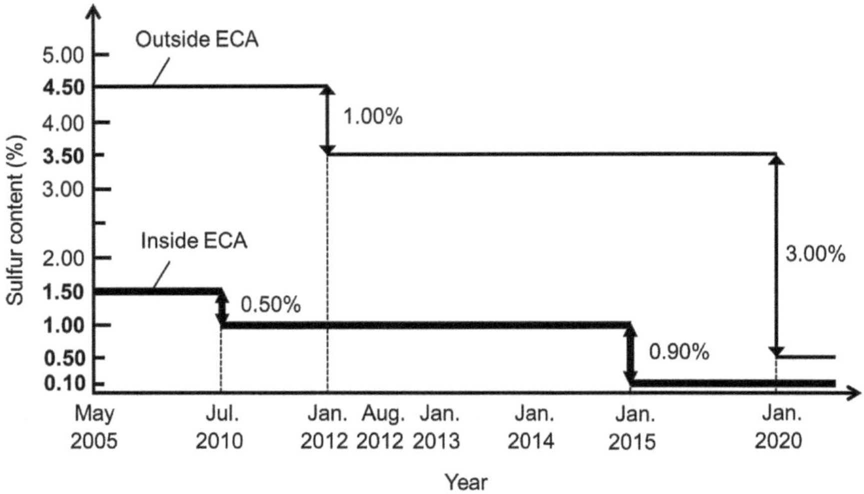

Figure 2.3.2 Sulfur content in fuel controlled outside and inside the emission control area (ECA) for marine diesel engines [7].

that is, outside and inside the ECA [7]. ECAs for SO_x regulation were determined as follows: (1) the Baltic Sea area, (2) the North Sea area, (3) the North American area, and (4) the Caribbean Sea area. As shown in Fig. 2.3.2, the regulation on sulfur content in fuel is controlled outside and inside the ECA for marine diesel engines. Outside the ECA, the limits for sulfur content, which are expressed in percent by mass, are regulated over time from 4.50% down to 3.50% (1.00% decrease) in January 1, 2012, and down to 0.50% (3.00% decrease) in January 1, 2020. Inside the ECA, the limits are regulated from 1.50% down to 1.00% (0.50% decrease) in July 1, 2010, and from 1.00% to 0.10% (0.90% decrease) in January 1, 2015.

As an alternative method, instead of utilizing approved fuels, an exhaust gas cleaning system (EGCS) such as a SO_x scrubber can be an approved subject to periodic parameter and emission checks, or the system may be equipped with a continuous emission monitoring system based on the 2009 Guidelines For Exhaust Gas Cleaning Systems as a resolution by Marine Environment Protection Committee. In this manner, the SO_2 (ppm)/CO_2 (%) ratio method is used to simplify the monitoring of SO_x emission and facilitates the approval of an EGCS where SO_2 (ppm) and CO_2 (volume%) are SO_2 and CO_2 concentrations in the exhaust gas, respectively. Table 2.3.1 shows the relationship between the limits of sulfur content in MARPOL

Table 2.3.1 Limits of sulfur content in Regulation 14 and the corresponding SO_2 (ppm)/CO_2 (volume%) ratio for facilitating approval of an exhaust gas cleaning system [8].

Limits of sulfur content mass% in Regulation 14	Corresponding SO_2 ppm/CO_2 volume% ratio
4.50	195.0
3.50	151.7
1.50	65.0
1.00	43.3
0.50	21.7
0.10	4.3

convention Annex VI, Regulation 14 and the corresponding SO_2 (ppm)/CO_2 (volume%) ratio. It is noted that the values in the left column are equivalent to those in the right column.

2.4 Particulate matter regulations

2.4.1 Particulate matter and related environmental issues

PM is a solid particle (particulate) or liquid droplet. PM in the atmosphere is called an aerosol, and PM itself often refers to aerosol. Generally, suspended particulate matter corresponds to particles having a diameter of 10 μm or less suspended in the atmosphere. Sources of PM can be natural or anthropogenic. Natural sources include volcanic ash, pollen, yellow sand (Asian Dust), sea salt, etc. Anthropogenic sources include soot, smog, fly ash, chemical mist, etc. In particular, PM due to anthropogenic sources is regarded as a problem. PM suspended in the atmosphere in large concentrations is considered as air pollution. In recent years, air pollution due to $PM_{2.5}$ has been recognized as a serious environmental issue. The term "PM" is recently often used and heard frequently. It is used as PM_d for PM with an aerodynamic diameter of d μm or less; for instance, $PM_{2.5}$ for PM with the diameter of 2.5 μm or less. Diesel emission contains PM. PM from diesel emission primarily consists of dry soot (carbon, C), soluble organic fraction, HC compounds, sulfate, etc. The amount of emitted PM tends to increase as combustion temperature decreases.

In many cases, inhaling PM can negatively affect human health because these are easily deposited in the respiratory tract and lungs. The typical range of PM sizes deposited in the path from the nose to the lungs are as follows: 7.0–11 μm in the nose, 4.7–7.0 μm in the pharynx, 3.3–4.7 μm in the trachea, 1.1–3.3 μm in the bronchial tubes, and less than 1.1 μm in the alveolus inside the lung. The deposited PM cannot be excreted immediately by the clearance mechanism of the human body and remains in place for a long time. Respiratory diseases are caused by PM deposition, especially in the alveolus. In the case where PM is a carcinogenic substance, the tissue could become cancerous. The effect of PM on human health will be explained in Chapter 5.

2.4.2 Particulate matter regulations for automobile engines

PM emission from automobiles is regarded as a serious problem. Therefore, similar to NO_x, regulations for PM emissions have been imposed regionally and worldwide. Fig. 2.4.1 shows typical diesel automobile exhaust PN in a solid curve and mass-weighted size distributions in a dotted curve [9]. The vertical axis expresses the normalized concentration $dC/C_{total}/d\log d_p$ and the horizontal axis expresses the particle diameter d_p, where C is the size dependent concentration, C_{total} is the total concentration, and $\log d_p$ is the common logarithm of d_p. Generally, it is a known that log-normal distribution tends to be the best fit for the single source aerosols. Therefore, the graph is drawn as a semi-log plot. Because the interval $d\log d_p$ between $\log d_p$ and $\log d_p + d\log d_p$ on the horizontal axis

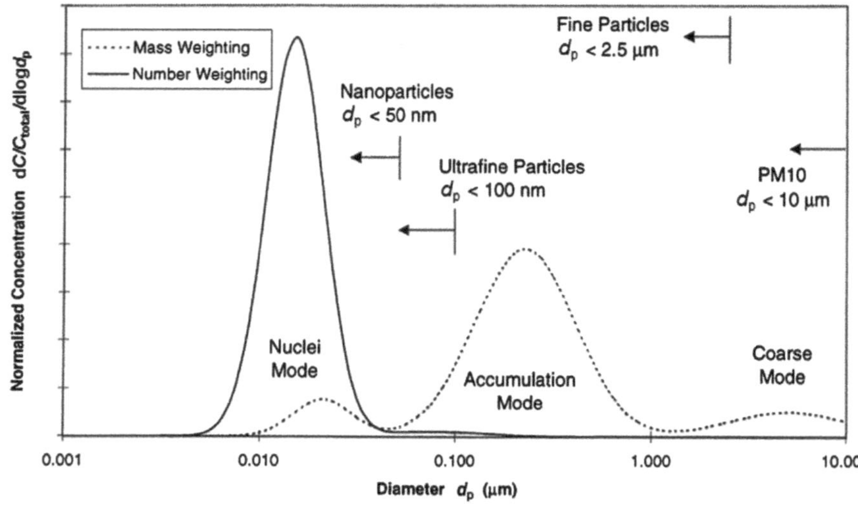

Figure 2.4.1 Typical diesel exhaust particle number- and mass-weighting size distributions [9].

differs depending on the measurement device, it is proper to divide dC by dlogd_p and C_{total}, and take it on the vertical axis as a normalized value. Consequently, regular data that can be compared among devices are obtained. Formations of PM are classified into three modes, that is, nuclei, accumulation, and coarse modes. In the nuclei mode, ultrafine combustion particles are produced by homogeneous and heterogeneous nucleation processes, and particles are formed in the atmosphere by gas-to-particle conversion. In the accumulation mode, nuclei-mode particles are coagulated, or vapors are condensed onto existing particles. Larger particles than those in the nuclei mode are formed by the processes in the accumulation mode. The coarse mode induces particles with a diameter larger than 1.0 μm. These coarse particles are scarcely formed during the diesel combustion process. They are rather formed through deposition and subsequent reentrainment of particulate material from walls of the combustion chamber and exhaust system. Most of the particle mass exists in the accumulation mode where the range of particle sizes is 0.1−0.3 μm, but most numbers of particles exist in the nuclei mode, where the range of particle sizes is 0.005−0.05 μm = 5−50 nm. In general, PM regulations focus on the mass of PM in emission. As a result, the target is PM with a size greater than 0.1 μm in the accumulation mode because they are dominant in terms of mass. In other words, PM regulations target relatively large PM. However, nanosized PM is regarded as a serious problem for human health. Nanosized PM in diesel emissions exist in the nuclei mode, and their number is much higher than those in the accumulation mode. Therefore, the total number of particles should be reduced, especially nanosized PM in emissions. This type of control is called PN regulation. Gasoline engines are not exempted from the issues of PM and larger PN given by the recent popularity of gasoline direct-injection engines, whose PM emissions are more difficult to reduce compared with conventional premixed combustion gasoline engines, from a technological point of view. PN regulation has been introduced as Euro 5b in 2011. PN regulations will be introduced worldwide from now on. The significant vehicle emission controls in Japan and in Europe are reviewed in this subsection.

In recent years, the following regulations were issued in Japan: the Long-Term Regulation was issued in 2005 and Post-New Long-Term Regulation was issued in 2009 for PM, NO_x, CO, and NMHC emissions, as mentioned in Subsection 2.2.2. Concerning with PM, in the case of vehicles weighing

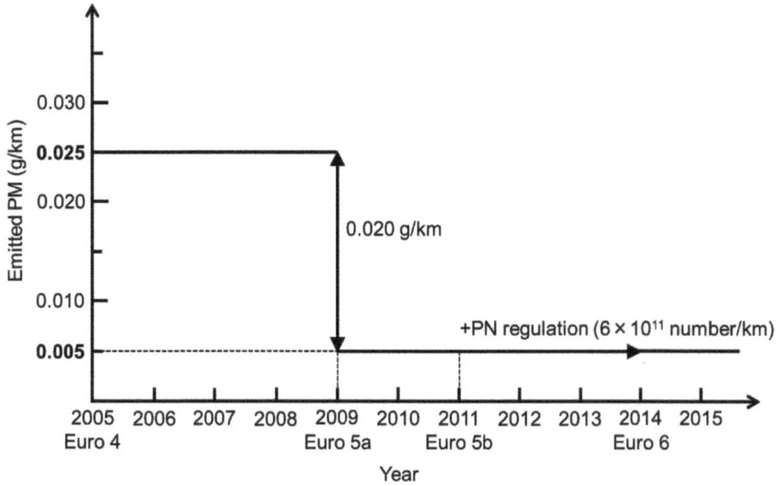

Figure 2.4.2 Recent particulate matter regulation in Europe for diesel passenger cars [4].

more than 3.5 ton, the regulation values are 0.027 g/kWh and 0.01 g/kWh for the Long-Term Regulation and Post-new Long-Term Regulation, respectively. It is noted that the values are represented by the ratio of PM mass from a target engine to output energy during a test cycle. As a result, PM has to be reduced by 0.027−0.01 = 0.017 g/kWh from 2005 to 2009. In addition, the regional regulation was also imposed from 2017.

Fig. 2.4.2 shows the recent PM regulation in Europe for diesel passenger cars. In Europe, as mentioned in Subsection 2.2.2, the exhaust emissions of new automobiles in the EU and EEA member states are controlled by European emission standards typically referred to as Euro 1, Euro 2, Euro 3, Euro 4, Euro 5, and Euro 6 for diesel passenger cars since 1992. Concerning with PM, in recent regulations for diesel passenger cars, the values are regulated from 0.025 g/km in Euro 4 down to 0.005 g/km in Euro 5a, Euro 5b, and Euro 6 (0.020 g/km decrease). It is noted that the values are measured in g/km, that is, the ratio of the mass of emitted PM to mileage. In addition, PN regulation is simultaneously imposed in Euro 5b from 2011. The regulation values are 6×10^{11} number/km. Fig. 2.4.3 shows the recent PM regulation in Europe for trucks and buses. In recent regulations for trucks and buses, the values are regulated from 0.02 g/kWh in Euro IV in 2005 down to 0.01 g/kWh in Euro VI in 2013, which corresponds 0.01 g/kWh decrease. The values are measured in g/kWh, that is, the ratio of the mass of emitted PM to output power.

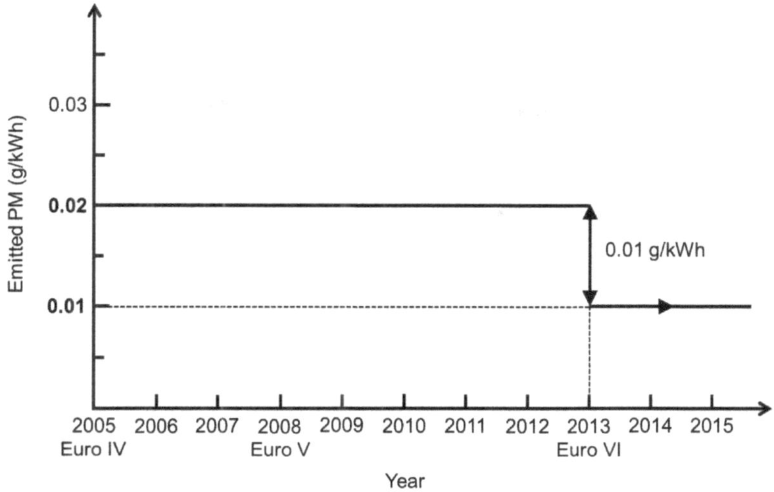

Figure 2.4.3 Recent particulate matter regulation in Europe for trucks and buses [5].

2.4.3 Particulate matter regulations for marine diesel engines

Worldwide regulations for emissions from marine diesel engines are managed by IMO. PM emissions from marine diesel engines are just as much of a problem as those from automobiles. The fuel of marine diesel engines such as MDO and MFO contains a larger amount of sulfur than fuel used in automobiles such as light fuel oil or GO. During the combustion process, the sulfur in fuel produces SO_x, i.e., SO_2 and sulfate, which is a type of PM, in emission. In a study [10] on the relationship between sulfur in marine fuels and PM emission, it was revealed that the PM conversion ratio from sulfur in fuel is 1.5–2.5% in a sulfur content range of 0.07–1.6%. The sulfate conversion ratio from sulfur in fuel is typically 5.0–13% in the sulfur content range. The experimental result shows that more than 50 mass% of PM consists of sulfate and its binding water. The ratio of binding water to sulfate is typically 1.3. In summary, the main PM component in marine diesel emission fueled by MFO or HSO is SO_x. Therefore, at present, instead of direct regulation of PM, the regulation of SO_x is imposed by IMO. The amount of SO_x in emission from ships is regulated by MARPOL convention Annex VI, Regulation 14 based on the sulfur content of fuel, as previously shown in Fig. 2.3.2. It is noted that the title of the regulation is Sulfur Oxides (SO_x) and Particulate Matter, which implies that the target is not only SO_x but also PM. The regulation sets a limit for the sulfur content of fuels. The values of the limit become more stringent in stages and different outside and inside the ECA, which are explained in Subsection 2.3.3.

2.5 Fuel regulations

2.5.1 Fuel and related environmental issues

Emission control technologies can be classified into fuel pretreatment, combustion improvement, and exhaust gas aftertreatment. Fuel pretreatment is the reduction of the causes that lead to the emergence of pollutants in fuel. An advantage of the pretreatment is that it is relatively effective and simple because it does not require additional devices with an engine. A disadvantage of the pretreatment is that the applicable pollutants are limited. An advantage of combustion improvement is that it does not require additional devices outside of the engine. A disadvantage of combustion improvement is that optimization and significant reduction are technically difficult to achieve. Exhaust gas aftertreatment is the reduction of pollutants in the exhaust gas. An advantage of the aftertreatment is that a large amount of pollutants can be reduced and the reduction can be applied to many kinds of pollutants. A disadvantage of the aftertreatment is that it requires additional devices outside an engine. In this book, combustion improvement and exhaust gas aftertreatment technologies are mainly treated and explained. However, pretreatment is of course important because it is a relatively simple and effective way to remove pollutants caused by contaminants in fuel. For example, SO_x in emission is caused by sulfur in fuel. If the sulfur were removed completely from the fuel, then SO_x emission from the engine could vanish without additional aftertreatment devices. To control this kind of emission, fuel components and their regulations are important and therefore treated in this subsection.

As alternative new fuels for diesel engines, biofuels are attracting attention. Biofuels, which are normally liquid or gas, are made from plants instead of fossil fuels. From the viewpoint of environmental protection or combatting with global warming, CO and CO_2 emitted from biofuel combustion are not regarded as greenhouse gases according to the concept of carbon neutral. Plants develop by photosynthesis while absorbing CO_2 in the atmosphere. Because the CO_2 emitted by an engine using biofuels originates from atmospheric CO_2, it can be considered that the total amount of CO_2 in the atmosphere does not change. Biofuels can be produced in a relatively short period; hence, they are regarded as renewable energy resources. Moreover, from the viewpoint of SO_x emission, biofuels contain little sulfur. Therefore, SO_x is not a regular component of biofuel emissions. One of the biofuels specified for diesel engines is called biodiesel fuel (BDF). BDF is produced from vegetable oil and/or animal fat by removing glycerol. There are several methods for producing BDF. In the conventional method, i.e., alkali catalyst method, vegetable oil and/or animal fat, from which moisture and free fatty acid are removed by pretreatment, is prepared. An appropriate amount of methanol (CH_3OH) and catalysts are added to the oil. Sodium hydroxide (NaOH) or potassium hydroxide (KOH) is typically used as a BDF catalyst. The vegetable oil and/or animal fat and the BDF catalyst are mixed at an appropriate temperature to induce transesterification. After mixing, the mixture is separated into fatty acid methyl ester (FAME) and glycerol by precipitation. Finally, BDF is obtained by removing glycerol and purification of FAME. In recent years, BDF standards have been also issued for its quality and utilization.

2.5.2 Fuel regulations and standards for automobile engines

In Japan, there are two quality standards for diesel automobile fuels. One is a mandatory standard called Law on the Quality Control of Gasoline and Other Fuels issued by the government. The other is a voluntary one called Japanese Industrial Standard (JIS) K 2204—Diesel Fuel, where K2204 is the number of the standard. The Law on the Quality Control of Gasoline and Other Fuels provides the requirements for components and fuel specifications [11]. The standard covers not only diesel fuel but also gasoline and kerosene fuels. Table 2.5.1 shows the current diesel fuel (light fuel oil or gas oil) regulation under the standard. In the table, although various properties are regulated, especially the cetane index, which is used as a substitute for the cetane number, indicates the ignition ability of diesel fuel. In the table, classification (objective) shows the purposes of the property for environmental protection or prevention of engine trouble. Fuels with higher values of cetane index ignite more easily. As for distillation, fuels with lower distillation temperatures are easier to evaporate.

BDF standards have been also issued for its quality and utilization. The Law on the Quality Control of Gasoline and Other Fuels provides requirements on a blend of BDF (5%) and diesel fuel (95%), which is termed B5. The quality standard for pure BDF, which is termed B100, is provided by the regulation.

Table 2.5.1 Diesel fuel (light fuel or gas oil) regulation in Japan under the Law on the Quality Control of Gasoline and Other Fuels [12].

Property	Standard	Classification (objective)
Sulfur	$\leq$ 0.001 mass%	Environment
Cetane index	$\geq$ 45	"
Distillation temperature for 90% distillation completed	$\leq$ 360°C	"
Triglyceride	$\leq$ 0.01 mass%	Prevention of engine trouble
Fatty acid methyl ester (FAME)	$\leq$ 0.1 mass% $\leq$ 5 mass%	"
[a]Methanol	$\leq$ 0.01 mass%	"
[a]Acid value	$\leq$ 0.13 mg(KOH)/g	"
[a]Total of formic, acetic, and propionic acids	$\leq$ 0.003 mass%	"
[b]Oxidative stability	$\geq$ 65 min	"

[a]In the case of FAME = 0.1–5 mass%.
[b]In the case of acid, value > 0.12 mg(KOH)/g.

In Europe, there is a quality standard for diesel fuels issued by the European Committee for Standardization, along with the European emission standards. Although the standard is not mandatory, the quality of fuels is observed by fuel suppliers in Europe. The standard primarily focuses on on-road (automobile) applications. Fuel providers add a marker or dye into fuels for off-road applications to distinguish it from fuels for on-road applications because the rate of taxation is lower than fuels for on-road applications. However, many European countries specify that the qualities of fuels for off-road applications are the same as those for on-road applications. Fuels for off-road applications can be used for on-road applications as well. In such cases, a marker or dye for taxation purposes is important. Other countries have a different standard for off-road applications. The sulfur, cetane number, and FAME contents of diesel fuels have been made more stringent in several steps. In 1993, the sulfur content limit was 0.2% (2000 ppm) and the cetane number was 49 for both on-road and off-road applications. This is referred to as Euro 1 diesel fuel. In 1996, the sulfur content was limited to 500 ppm and is referred to as Euro 2 diesel fuel. In 1999, the sulfur content and cetane number were limited to 350 ppm and 50, respectively (Euro 3). In 2006, the sulfur content was limited to 50 ppm (Euro 4). In 2009, the sulfur and FAME contents were limited to 10 ppm and 7% (Euro 5).

Table 2.5.2 Diesel fuel standard in Europe made based on EN 590:2009 [13].

Property	**Limits**	
	Min.	**Max.**
Cetane number	51.0	–
Cetane index	46.0	–
Density at 15°C	820.0 kg/m^3	845.0 kg/m^3
Polycyclic aromatic hydrocarbon	–	11 mass%
Sulfur content	–	10.0 mg/kg
Flash point	> 55°C	–
Carbon residue on 10% distillation residual	–	0.30 mass%
Ash content		0.01 mass%
Water content		200 mg/kg
Total contamination		24 mg/kg
Copper strip corrosion (3 h at 50°C)	Class 1	
Fatty acid methyl ester		7.0 volume%
Oxidation stability g/m^3	–	25 g/m^3
	20 h	–
Lubricity, wear scar diameter (wsd 1.4) at 60°C	–	460 μm
Viscosity at 40°C	2.00 mm^2/s	4.50 mm^2/s
Distillation recovered at 250°C		< 65 volume%
Distillation recovered at 350°C	85 volume%	
Distillation 95 volume% recovered at		360°C

Table 2.5.2 shows the diesel fuel standard in European Norm (EN) 590 established in 2009, and it is valid at the present. The numbers after EN are assigned to objectives, and EN 590 is for specification for automotive diesel. In the table, both the cetane index and cetane number indicate the self-ignition ability of diesel fuel. A higher value reduces the risk of diesel knocking. The cetane index is calculated based on the density and distillation range of the fuel. The cetane number is measured, but accurate measurements are rather difficult to obtain. Copper corrosion indicates metallic corrosion due to the sulfur content in the fuel. The lubricity is measured by a wear scar diameter (wsd) of a test ball in the high-frequency reciprocating rig test, which is a tribological measurement. The wsd depends on the vapor pressure in the test. In the table, the wear scar diameter (wsd 1.4) stands for the mean wsd normalized to be at a standard vapor pressure of 1.4 kPa in the test. The distillation is regulated by recovery rates at temperatures of 250°C and 350°C and a distillation temperature at a recovery rate of 95 volume%.

BDF standards have also been issued in Europe. The standard provides requirements for pure BDF called B100. The requirements for a blend of BDF and diesel fuel, such as B5 (5% BDF inclusion) and B7 (7% BDF inclusion), are provided by a standard that mentions that FAME should meet the requirements of the BDF standard.

2.5.3 Fuel regulations and standards for marine diesel engines

Petroleum fuels are mostly used for marine diesel engines. There are different grades of petroleum fuels. In general, high-grade fuels are environment-friendly, but the prices are high. In the case of a liquefied natural gas (LNG) tanker, the boil-off gas (BOG) of LNG is used as fuel. The BOG is natural evaporation gas from LNG in the tank, and it is an inevitable consequence of its storage. The advantage of LNG utilization is that SO_x, NO_x, and CO_2 emissions are greatly reduced. It is said that SO_x emission is almost zero, while NO_x and CO_2 can be reduced by approximately 80% and 25%, respectively. However, a disadvantage is that a large special tank is needed to store LNG, which creates problems in terms of cost and storage space in a ship. In view of the advantages of LNG and petroleum fuel, dual-fuel engines have been developed to alternately use LNG and petroleum fuel explained in Chapter 5.

Fuel regulations for marine diesel engines currently focus mainly on sulfur content in fuels. As mentioned in Subsection 2.3.3, the sulfur content in fuels is regulated by the MARPOL convention Annex VI, Regulation 14. The values of the limit become more stringent in stages and different outside and inside the ECA as explained in Subsection 2.3.3. In addition, other countries and regions such as the United States, Turkey, and China have their own regional regulations. In California, for instance, the California Air Resources Board (CARB) regulation is imposed.

In addition, fuel oil quality is regulated by the MARPOL convention Annex VI, Regulation 18. Regulation 18 mentions the requirements which the fuel oil delivered to and used onboard ships for combustion purposes should meet and that the fuel oil should be hydrocarbon blends derived from petroleum refining. Further explanations are presented in A.2 of Appendix.

In general, petroleum fuels for marine diesel engines are classified as marine gas oil (MGO), MDO, intermediate fuel oil (IFO), MFO, and heavy fuel oil. An international standard of fuels for marine diesel engines is provided and established in 2017 as ISO 8217 [14] by the International

Organization for Standardization (ISO). The scope of the ISO 8217:2017, which is the sixth edition of the specifications of fuels for marine diesel engines, mentions the requirements for fuels for use in marine diesel engines and boilers, prior to conventional onboard treatment such as settling, centrifuging, and filtration, before use. The specifications for fuels in this document can also be applied to fuels used in stationary diesel engines of the same or similar type as those used for marine purposes.

Table 2.5.3 shows a list on the standard for marine distillate fuels and distillate FAMEs making based on the ISO 8217:2017. In the table, the grades of the fuels are shown with abbreviations such as DMX, DMA, DMZ, and DMB for marine distillate fuels and DFA, DFZ, and DFB for distillate FAMEs. It is noted that DMX, DMA, and DFA are higher grades, and DMB and DFB are lower ones. In the table "Report" means a requirement of the report on the cloud point and cold filter plugging point in the winter for DMA/DFA and DMZ/DFZ. MGO and MDO correspond to DMA and DMB, respectively.

Table 2.5.3 Standard on marine distillate fuels made based on ISO 8217:2017 [14].

Parameter	Limit	DMX	DMA	DFA	DMZ	DFZ	DMB	DFB
Viscosity at 40°C mm^2/s	Max.	5.500	6.000		6.000		11.00	
	Min.	1.400	2.000		3.000		2.000	
Microcarbon residue at 10% residue mass%	Max.	0.30	0.30		0.30		–	
Density at 15°C kg/m^3	Max.	–	890.0		890.0		900.0	
Microcarbon residue mass%	Max.	–	–		–		0.30	
Sulfur mass%	Max.	1.00	1.00		1.00		1.50	
Water volume%	Max.	–	–		–		0.30	
Total sediment by hot filtration mass%	Max.	–	–		–		0.10	
Ash mass%	Max.	0.010	0.010		0.010		0.010	

Continued

Table 2.5.3 Standard on marine distillate fuels made based on ISO 8217:2017 [14].—cont'd

Parameter	Limit	DMX	DMA	DFA	DMZ	DFZ	DMB	DFB
Flash point °C	Min.	43.0	60.0		60.0		60.0	
Pour point in winter °C	Max.	–	−6		−6		0	
Pour point in summer °C	Max.	–	0		0		6	
Cloud point in winter °C	Max.	−16	Report		Report		–	
Cloud point in summer °C	Max.	−16	–		–		–	
Calculated cetane index	Min.	45	40		40		35	
Acid number mg(KOH)/g	Max.	0.5	0.5		0.5		0.5	
Oxidation stability g/m^3	Max.	25	25		25		25	
Fatty acid methyl ester (FAME)	Max.	–	–	7.0	–	7.0	–	7.0
Lubricity, corrected wear scar diameter (wsd 1.4) at 60°C μm	Max.	520	520		520		520	
Hydrogen sulfide mg/kg	Max.	2.00	2.00		2.00		2.00	
Appearance		Clear and bright					–	

Table 2.5.4 shows a list on the standard for other marine fuels or marine residual fuels making based on the ISO 8217:2017. In the table, the grades of the fuel are shown with abbreviations such as RMA, RMB, RMD, RME, RMG, and RMK. It is noted that RMA is the highest grade and

Table 2.5.4 Standard for marine residual fuels made based on ISO 8217:2017 [14].

Parameter	Limit	RMA 10	RMB 30	RMD 80	RME 180	RMG 180	RMG 380	RMG 500	RMG 700	RMK 380	RMK 500	RMK 700
Kinematic viscosity at 50°C, mm^2/s	Max.	10.00	30.00	80.00	180.0	180.0	380.0	500.0	700.0	380.0	500.0	700.0
Density at 15°C, kg/m^3	Max.	920.0	960.0	975.0	991.0	991.0				1010.0		
MicroCarbon residue, mass%	Max.	2.50	10.00	14.00	15.00	18.00				20.00		
Aluminum + Silicon, mg/kg	Max.	25	40		50	60						
Sodium, mg/kg	Max.	50	100		50	100						
Ash, mass%	Max.	0.040	0.070			0.100				0.150		
Vanadium, mg/kg	Max.	50	150			350				450		
Calculated carbon aromaticity index	Max.	850	860			870						
Water, volume%	Max.	0.30	0.50									
Pour point (upper) in summer, °C	Max.	6		30								
Pour point (upper) in winter, °C	Max.	0		30								
Flash point, °C	Min.	60.0										
Sulfur, mass%	Max.	To comply with statutory requirements as defined by purchaser.										
Total sediment, aged, mass%	Max.	0.10										
Acid number, mg(KOH)/g	Max.	2.5										
Hydrogen sulfide, mg/kg	Max.	2.00										

RMK is the lowest one. In the table, for instance, RMA 10 means RMA grade with a kinematic viscosity of 10.00 mm^2/s at 50°C. In the table, CCAI stands for the calculated carbon aromaticity index (a unitless quantity) defined by an equation: $CCAI = \rho - 140.7\log\log(\mu + 0.85) - 80.6$, which is calculated using the fuel density ρ at 15°C and viscosity μ at 50°C. A higher CCAI indicates that the ignition delay will increase in the diesel engine. An appropriate guidance on fuel treatment systems for diesel engines is published by the International Council on Combustion Engines (CIMAC) [15]. It is noted that the requirements for gas turbine fuels used in marine applications are specified in ISO 4261:2013, which is established in 2013.

References

[1] Japan Ozone Kyokai, Ozone Handbook, Sanyu Syobou Co., 1999, pp. 9–45 (in Japanese).

[2] Ministry of Land, Infrastructure, Transport and Tourism, Post shin-chokikisei no gaiyou (Outline of post-new long-term regulations), 2018, http://www.mlit.go.jp/kisha/kisha08/09/090325/01.pdf.

[3] Ministry of Land, Infrastructure, Transport and Tourism, Doro yusou sharyo no annzenkijun, doro yusou sharyo no hoankijun no saimoku wo sadameru kokuji tou no ichibu kaisei (Partial revision of safety standard on road vehicles and its announcement of the details), 2018, http://www.mlit.go.jp/common/001094623.pdf.

[4] DieselNet, Emission standards: EU: Cars and light trucks, 2018, https://www.dieselnet.com/standards/eu/ld.php.

[5] DieselNet, Emission standards: EU: Heavy-duty truck and bus engines, 2018, https://www.dieselnet.com/standards/eu/hd.php.

[6] M. Okubo, T. Kuwahara, K. Yoshida, M. Kawai, K. Hanamoto, K. Sato, T. Kuroki, Total marine diesel emission control technology using nonthermal plasma hybrid process, in: Proceedings of the 27th CIMAC World Congress on Combustion Engine Technology (CIMAC 2013), Paper No.153 (14 pages), 2013.

[7] J. Herdzik, Emissions from marine engines versus IMO certification and requirements of tier 3, J. KONES Powertrain Transp. 18 (2) (2011) 161–167.

[8] International Maritime Organization (IMO), Guidelines for Exhaust Gas Cleaning Systems, 2009, http://www.imo.org/en/KnowledgeCentre/IndexofIMOResolutions/Marine-Environment-Protection-Committee-(MEPC)/Documents/MEPC.184(59).pdf.

[9] D.B. Kittelson, Engines and nanoparticles: a review, J. Aerosol Sci. 29 (5–6) (1998) 575–588.

[10] A. Ohashi, M. Ikame, E. Ishimura, S. Nishio, M. Takagi, Z. Xu, Evaluation of collection loss of sulfate in exhaust gas dilution system, J. Japan Inst. Mar. Eng. 45 (3) (2010) 415–420 (in Japanese).

[11] Agency for Natural Resources and Energy, Sekiyu Seihinn No Hinshitsu Kakuho Ni Tsuite (Law on the Quality Control of Gasoline and Other Fuels), 2014, http://www.enecho.meti.go.jp/category/resources_and_fuel/distribution/hinnkakuhou/.

[12] Agency for Natural Resources and Energy, Sekiyu seihin no hinshitsu kakuho nituite (Law on the quality control of gasoline and other fuels), 2018, http://www.enecho.meti.go.jp/category/resources_and_fuel/distribution/hinnkakuhou/.

[13] European Committee for Standardization, Automotive fuels – diesel – requirements and test methods, 2018. Ref. No. EN 590:2009: E, http://www.envirochem.hu/www.envirochem.hu/documents/EN_590_2009_hhV05.pdf.
[14] International Organization for Standardization, 2018. ISO 8217:2017, https://www.iso.org/standard/64247.html.
[15] International Council on Combustion Engines (CIMAC), Recommendation No. 25 Recommendations Concerning the Design of Heavy Fuel Treatment Plants for Diesel Engines, May 2006.

CHAPTER 3

Principle and design of emission control systems

Contents

New Technologies for Emission Control in Marine Diesel Engines
ISBN 978-0-12-812307-2
https://doi.org/10.1016/B978-0-12-812307-2.00003-1

3.1 Air pollutants generated in exhaust gas

In this section, the generation of air pollutants in the exhaust gas from marine diesel engine such as nitrogen oxide (NO_x), sulfur oxide (SO_x) [1,2], particulate matter (PM), and carbon dioxide (CO_2) are described. In order to know the combustion state roughly, smoke color is important.

Smoke color observed represents the state of smoke emitted from stack, and the combustion state can be qualitatively diagnosed from it. The meaning of smoke color is explained as follows. Blue-colored smoke shows that lubrication oil and/or engine oil is burning. Yellow-colored smoke shows that sulfur content is burning to induce SO_x in the exhaust gas. White-colored smoke shows that water vapor is included in the smoke. Black-colored smoke shows that excess fuel is burning with lower air-to-fuel ratio, and incomplete burning is occurring. Brown-colored smoke shows that NO_x such as NO_2 is included in the smoke. Because, for the fuel combustion process, the amount of gaseous components or byproducts generated can be estimated quantitatively by a combustion calculation, the procedure for it is explained in the following.

3.1.1 Combustion calculation for liquid and solid fuels

To design an exhaust gas aftertreatment system, a combustion calculation is necessary to estimate the flow rate of the exhaust gas and gaseous components or byproducts generated. When calculating the combustion of liquid and solid fuels, it is useful to do so based on 1 kg of fuel. The amount of air required for the combustion of 1 kg of fuel in Nm^3/kg, where N means the standard state (0°C and 101.325 kPa), and the amount of combustion exhaust gas in Nm^3/kg should be obtained. The composition of liquid/solid fuel is generally given as a mass ratio. Herein, the carbon, hydrogen, sulfur, and nitrogen contents in 1 kg of fuel are represented by c, h, s, and n kg, respectively. When liquid or solid fuel is combusted, CO_2, CO, H_2O, NO, NO_2, SO_2, and SO_3 are mainly generated in the exhaust gas. It is assumed in the following calculation that the concentrations of CO_2, H_2O, NO, and SO_2 are higher as byproducts, and they are considered from the combustion oxidation of C, H, N, and S. The procedure used to conduct the combustion calculation is presented [3]. It is noted that the atomic weights are assumed to be H = 1.01 kg/kmol, C = 12.01 kg/kmol, N = 14.01 kg/kmol, O = 16.00 kg/kmol, and S = 32.06 kg/kmol, and the air composition ratio is N_2: O_2 = 79: 21.

The theoretical volume of oxygen required for 1 kg of fuel O_0 Nm^3/kg should be calculated. It is noted that calculation based on Nm^3 is carried out here although that based on kmol is also possible. The main combustion reactions for each composition based on the mass or volume of each component are as follows.

Carbon:

$$\begin{aligned} &C + O_2 = CO_2 \\ &12.01\ \text{kg}: 22.4\ \text{Nm}^3: 22.4\ \text{Nm}^3 \\ &c\ \text{kg}: 22.4\ c/12.01\ \text{Nm}^3: 22.4\ c/12.01\ \text{Nm}^3 \\ &c\ \text{kg}: 1.865\ c\ \text{Nm}^3: 1.865\ c\ \text{Nm}^3 \end{aligned} \tag{3.1}$$

For carbon of c kg, $22.4c/12.01 = 1.865c\ \text{Nm}^3$ of oxygen is required, and the same volume of $1.865c\ \text{Nm}^3$ of carbon dioxide is generated. Similar calculation can be achieved for hydrogen, sulfur, and nitrogen.

Hydrogen:

$$\begin{aligned} &H + O_2/4 = H_2O/2 \\ &1.01\ \text{kg}: 22.4/4\ \text{Nm}^3: 22.4/2\ \text{Nm}^3 \\ &h\ \text{kg}: 22.4\ h/4.04\ \text{Nm}^3: 22.4\ h/2.02\ \text{Nm}^3 \\ &h\ \text{kg}: 5.545\ h\ \text{Nm}^3: 11.089\ h\ \text{Nm}^3 \end{aligned} \tag{3.2}$$

Sulfur:

$$\begin{aligned} &S + O_2 = SO_2 \\ &32.06\ \text{kg}: 22.4\ \text{Nm}^3: 22.4\ \text{Nm}^3 \\ &s\ \text{kg}: 22.4\ s/32.06\ \text{Nm}^3: 22.4\ s/32.06\ \text{Nm}^3 \\ &s\ \text{kg}: 0.699\ s\ \text{Nm}^3: 0.699\ s\ \text{Nm}^3 \end{aligned} \tag{3.3}$$

Nitrogen:

$$\begin{aligned} &N + O_2/2 = NO \\ &14.01\ \text{kg}: 22.4/2\ \text{Nm}^3: 22.4\ \text{Nm}^3 \\ &n\ \text{kg}: 22.4\ n/28.02\ \text{Nm}^3: 22.4\ n/14.01\ \text{Nm}^3 \\ &n\ \text{kg}: 0.799\ n\ \text{Nm}^3: 1.599\ n\ \text{Nm}^3 \end{aligned} \tag{3.4}$$

Therefore, the theoretical oxygen volume required for 1 kg of fuel is

$$O_0(\text{Nm}^3/\text{kg}) = 1.865\ c + 5.545\ h + 0.699\ s + 0.799\ n \tag{3.5}$$

The theoretical air volume A_0 Nm^3/kg for 1 kg of fuel is given because the air contains approximately 21% oxygen.

$$A_0 = (1.865\ c + 5.545\ h + 0.699\ s + 0.799\ n)/0.21 \tag{3.6}$$

Required air volume A Nm^3/kg for 1 kg of fuel is usually larger than the theoretical air volume in actual diffusion combustion equipment.

$$A = \lambda A_0 \tag{3.7}$$

where λ is the ratio of the actual air volume to the theoretical air volume, which is called the air ratio.

The volume of combustion exhaust gas for 1 kg of fuel is calculated by the sum of the fuel volume, air volume, and volume ratio before and after combustion. In the case of liquid and solid fuel, the volume of the fuel is negligibly small compared with the combustion air. The amount of air is calculated using Eq. (3.7). From the combustion reactions of Eqs. (3.1) and (3.3), the volume difference before and after combustion is zero (no change in volume) for carbon and sulfur. For hydrogen and nitrogen, based on reactions (3.2) and (3.4), the differences exist. It is noted that the definition of wet combustion exhaust gas is a combustion gas containing all moisture generated through the combustion of hydrogen as water vapor. On the other hand, the definition of dry combustion exhaust gas is a combustion gas excluding all moisture vapor condensed into liquid water. The total volume increase is $5.545h + 0.799n$ Nm^3/kg in the wet combustion exhaust gas and $-5.545h + 0.799n$ Nm^3/kg in the dry combustion exhaust gas, respectively. Therefore, the volume of the wet combustion exhaust gas G Nm^3per 1 kg fuel is

$$G = \lambda A_0 + 5.545\,h + 0.799\,n \tag{3.8}$$

The volume of the dry combustion gas G' Nm^3 per 1 kg fuel is

$$G' = \lambda A_0 - 5.545\,h + 0.799\,n \tag{3.9}$$

Generally, because $n << h$ is hold, $0.799\,n$ of the final terms on the right-hand side of Eqs. (3.8) and (3.9) can be usually neglected.

CO_2 concentration in the dry combustion exhaust gas (volume $= G'$) is calculated because a total of $1.865c$ Nm^3 of CO_2 is produced from carbon of c kg from reaction (3.1).

$$CO_2 = \frac{1.865c}{G'} \times 100\% \tag{3.10}$$

SO_2 concentration in the dry combustion gas (volume $= G'$) is calculated because a total of $0.699s$ Nm^3 of SO_2 is generated from sulfur of s kg from reaction (3.3).

$$SO_2 = \frac{0.699s}{G'} \times 10^6 \text{ ppm} \tag{3.11}$$

Fuel NO concentration in the dry combustion gas (volume = G') is calculated because a total of 1.599 n Nm^3 is generated from nitrogen n kg from reaction (3.4).

$$NO = \frac{1.599n}{G'} \times 10^6 \text{ ppm} \tag{3.12}$$

It should be noted that the efficiency of converting the nitrogen content in fuel oil to fuel NO is actually 40–70%, and not all nitrogen is usually converted into NO. Furthermore, thermal NO generated through the oxidation of N_2 in the air cannot be calculated in the present procedure. It can be calculated with the procedure in Subsection 3.1.3.

Example 1: How much air (Nm^3/kg) is required to completely burn heavy oil, the components of which are 86.5 mass% carbon, 12.0 mass% hydrogen, 1.4 mass% sulfur, and 0.1 mass% nitrogen at an air ratio of 1.2? Further, calculate the concentration of CO_2 in the exhaust gas.

Solution 1: The theoretical air volume is calculated from Eq. (3.6).

$$\begin{aligned} A_0 &= (1.865 \times 0.865 + 5.545 \times 0.120 + 0.699 \times 0.014 + 0.799 \\ &\quad \times 0.001)/0.21 \\ &= 10.901 \text{ Nm}^3/\text{kg} \end{aligned}$$

The actual required air volume is calculated from Eq. (3.7).

$$A = \lambda A_0 = 1.2 \times 10.901 = 13.08 \text{ Nm}^3/\text{kg}$$

The volume of the dry combustion gas G' Nm^3 per 1 kg of fuel is

$$\begin{aligned} G' &= \lambda A_0 - 5.545h + 0.799n = 13.08 - 5.545 \times 0.120 + 0.799 \times 0.001 \\ &= 12.42 \text{ Nm}^3/\text{kg} \end{aligned}$$

The CO_2 concentration is calculated from Eq. (3.10) as

$$CO_2 = \frac{1.865 \times 0.865}{12.42} \times 100\% = 13.0\% \quad \text{(Answer)}$$

Example 2: Heavy oil having a composition of 86% carbon by mass, 12% hydrogen, and 2% sulfur is completely burned at an air ratio of 1.2 using a boiler with a decomposition of sulfur oxide (De-SO_x) apparatus.

Calculate the desulfurization efficiency % when the SO_2 concentration in the dry combustion gas is 68 ppm after the treatment.

Answer 2: For 1 kg of heavy oil, C of 0.86 kg, H of 0.12 kg, and S of 0.02 kg are included. The O_2 required by the combustion of C, H, and S for 1 kg of fuel is as follows:

C: $x = 22.4 \times 0.86/12.01 = 1.604\ \text{Nm}^3$

H: $x = 22.4 \times 0.12/4.04 = 0.665\ \text{Nm}^3$

S: $x = 22.4 \times 0.02/32.06 = 0.0140\ \text{Nm}^3$

A total O_2 of $1.604 + 0.665 + 0.0140 = 2.283\ \text{Nm}^3$ is required. The value is also obtained from Eq. (3.5) with $n=0$.

Air volume is $A_0 = 2.283/0.21 = 10.87\ \text{Nm}^3$ and $N_2 = 0.79 \times 10.87 = 8.587\ \text{Nm}^3$. Inside the dry combustion gas, CO_2 of 1.604 Nm^3, N_2 of 8.587 Nm^3, excess air of $(\lambda-1)A_0 = 10.87 \times 0.2 = 2.174\ \text{Nm}^3$, and SO_2 of 0.0140 Nm^3 are included. Therefore, raw SO_2 concentration before De-SO_x treatment and De-SO_x efficiency are calculated as

$$SO_2 = \frac{0.0140}{1.604 + 8.587 + 2.174 + 0.0140} \times 10^6\ \text{ppm}$$
$$= 1130\ \text{ppm} \quad \text{(Answer)}$$

$$\text{De} - SO_x\ \text{efficiency} = \left(1 - \frac{68}{1130}\right) \times 100 = 94.0\% \quad \text{(Answer)}$$

Example 3: Cetane ($C_{16}H_{34}$) is a typical component for oil fuel of marine diesel engines. What is the theoretical volume of oxygen (Nm^3) required to burn 1 kg of cetane and CO_2 concentration in the wet combustion exhaust gas when the air ratio λ is 1.2?

Answer 3: The burning equation of 1 kg cetane is

$$C_{16}H_{34} + 24.5O_2 = 16CO_2 + 17H_2O$$

1 kmol: 24.5 kmol: 16 kmol: 17 kmol

226.5 kg: $24.5 \times 22.4\ \text{Nm}^3$: $16 \times 22.4\ \text{Nm}^3$: $17 \times 22.4\ \text{Nm}^3$

1 kg: $24.5 \times 22.4/226.5\ \text{Nm}^3$: $16 \times 22.4/226.5\ \text{Nm}^3$: $17 \times 22.4/226.5\ \text{Nm}^3$.

Based on the proportional relationship of this combustion formula, the amount of oxygen required for combustion is $24.5 \times 22.4\ \text{Nm}^3$ of oxygen with the combustion of 226.5 kg of cetane. Therefore, for 1 kg of cetane,

$$O_0 = 24.5 \times 22.4/226.5 = 2.423\ \text{Nm}^3 \quad \text{(Answer)}$$

The theoretical air volume is $A_0 = 2.423/0.21 = 11.54\ \mathrm{Nm}^3$. The volume of combustion exhaust gas is calculated by the sum of the air volume and volume difference before and after combustion. Because the volume after the combustion is $(16 + 17) \times 22.4/226.5 = 3.264\ \mathrm{Nm}^3$ and the volume before the combustion is $2.423\mathrm{Nm}^3$, the wet combustion gas volume for 1 kg cetane is

$$\begin{aligned} G' &= \lambda A_0 + (\text{the volume difference before and after the combustion}) \\ &= 1.2 \times 11.54 + (3.264 - 2.423) = 14.69\ \mathrm{Nm}^3 \end{aligned} \tag{3.13}$$

The CO_2 gas volume is $16 \times 22.4/226.5 = 1.582\ \mathrm{Nm}^3$. Therefore, CO_2 concentration is

$$CO_2 = \frac{1.582}{14.69} \times 100 = 10.8\% \quad (\text{Answer})$$

It is known from this result that the concentration of CO_2 in the combustion of marine-diesel-engine oil fuel is typically 11%.

3.1.2 Combustion calculation for gas fuel

For the combustion calculation of gaseous fuel, it is convenient to consider $1\ \mathrm{Nm}^3$ of gaseous fuel where N means the standard state (0°C and 101.325 kPa) [3]. The calculation can be carried out while assuming an ideal gas law ($1\ \mathrm{kmol} = 22.4\ \mathrm{Nm}^3$). Taking methane ($CH_4$) as an example, which is a typical gas fuel for diesel engines, the calculation method is as shown below.

Theoretical volume of oxygen $O_0\ \mathrm{Nm}^3$ required for the combustion of gaseous fuel should be first determined. The combustion reaction formula of the fuel molecules are

$$\begin{gathered} CH_4 + 2O_2 = CO_2 + 2H_2O \\ 1\ \mathrm{Nm}^3\!: 2\ \mathrm{Nm}^3\!: 1\ \mathrm{Nm}^3\!: 2\ \mathrm{Nm}^3 \end{gathered} \tag{3.14}$$

From this combustion equation, it can be seen that $O_0 = 2\ \mathrm{Nm}^3$ of oxygen is required for the combustion of $1\ \mathrm{Nm}^3$ of methane.

The theoretical air volume A_0 equals $O_0/0.21$ because the oxygen concentration in air is 21 volume%. In the case of methane, because $O_0 = 2\ \mathrm{Nm}^3$, $A_0 = 2/0.21 = 9.52\ \mathrm{Nm}^3$.

The required air volume A equals λA_0 because, for diffusion combustion of engines, the air quantity supplied is generally greater than the theoretical amount in many cases to achieve complete combustion, where λ (normally $\lambda > 1$) is the air ratio or air equivalence ratio. In the case of

$\lambda = 1.2$, this is also called an excess air ratio of 20%. The required airflow at an air ratio of 1.2 is $A = 1.2 \times 9.52 = 11.42$ Nm^3.

The volume of combustion exhaust gas can be expressed in two ways: the volume of the wet combustion exhaust gas G Nm^3 and the volume of dry combustion exhaust gas G' Nm^3. The volume of wet combustion exhaust gas is the gas volume including water vapor in the exhaust gas or without any condensation of water vapor, and the volume of dry combustion exhaust gas is the volume of gas excluding water vapor or with complete condensation of all water vapor. These values are calculated from combustion reaction equations. The combustion gas volume can be calculated by the sum of the fuel volume, the combustion air volume, and the volume difference after and before the combustion. Taking methane as an example again, for the fuel volume of 1 Nm^3, the combustion air volume is λA_0 and the volume difference after and before the combustion is $(1 + 2) - (1 + 2) = 0$ Nm^3 from reaction (3.14) when considering the wet combustion gas. Therefore, the wet combustion gas volume G Nm^3 for the combustion of 1 Nm^3 methane is

$$G = 1 + \lambda A_0 \tag{3.15}$$

Similarly, when considering the dry combustion gas, the volume difference after and before the combustion is $1 - (1 + 2) = -2$ Nm^3. Therefore, the dry combustion gas volume G' Nm^3 for the combustion of 1 Nm^3 methane is

$$G' = 1 + \lambda A_0 - 2 \tag{3.16}$$

When methane is burned at an air ratio of 1.2, $G = 1 + 1.2 \times 9.52 = 12.42$ Nm^3 and $G' = 1 + 1.2 \times 9.52 - 2 = 10.42$ Nm^3.

The concentration of carbon dioxide for methane combustion is calculated as follows. As can be seen from the combustion reaction, 1 Nm^3 of carbon dioxide is produced per 1 Nm^3 of methane. The CO_2 concentration (%) per dry combustion gas volume G' is given by the following equation.

$$CO_2 = \frac{1}{G'} \times 100\% \tag{3.17}$$

Example 1: What is the theoretical volume of oxygen required to burn 1 kg of methane and CO_2 concentration in the dry combustion gas with the air ratio of 1.2?

Answer 1: 1 kg methane corresponds to $22.4/16.05 = 1.396$ Nm^3 methane. Therefore, the theoretical volume of oxygen is

$$O_2 = 2 \times 1.396 = 2.792 \text{ Nm}^3 \quad \text{(Answer)}$$

The theoretical CO_2 concentration is calculated from Eq. (3.17) when the air ratio is 1.2 as follows.

$A_0 = 2.792/0.21 = 13.30\ \mathrm{Nm}^3$, $G' = 1 + 1.2 \times 13.30 - 2 = 14.96\ \mathrm{Nm}^3$

$$CO_2 = \frac{1}{14.96} \times 100\% = 6.68\ \% \quad \text{(Answer)}$$

It is known from this result that the concentration of CO_2 in the gas fuel combustion is typically about 6.7%.

3.1.3 NO_x, SO_x, and particulate matter formations by combustion

NO_x, SO_x, and PM formations by combustion are described based on Ref. [1]. NO_x is emitted into the air during combustion. NO_x emissions are mainly represented by NO (nitrogen monoxide) + NO_2 (nitrogen dioxides). Fig. 3.1.1 shows the classification of NO generated in combustion. NO is dominant in the combustion process and classified into fuel NO and thermal NO. Fuel NO is induced by the combustion of nitrogen in the fuel. The formation of fuel NO proceeds radically at the air excess ratio of $\lambda > 1.0$ and thus, can be reduced at $\lambda < 1.0$. Fuel NO formation is also affected by the mixture of fuel and air or O_2 and is enhanced in a well-mixed environment. In diffusive mixings, turbulent diffusion generates a larger amount of fuel NO than laminar diffusion. Although the generated amount of fuel NO depends on the amount of fuel nitrogen (fuel N) and increases with increase in the fuel N, the conversion ratio from fuel N to NO usually becomes smaller as the amount of fuel N becomes larger. To reduce the fuel NO in the exhaust gas, there are several methods. They are to remove fuel N in pretreatment, to decrease O_2 in combustion, and to suppress the mixing of fuel and air (O_2).

In heavy oil, fuel N is present as pyridine, quinoline, and pyrrole in the fuel. These compounds react with oxygen during combustion to generate fuel NO. Therefore, fuel NO is usually generated during the combustion of

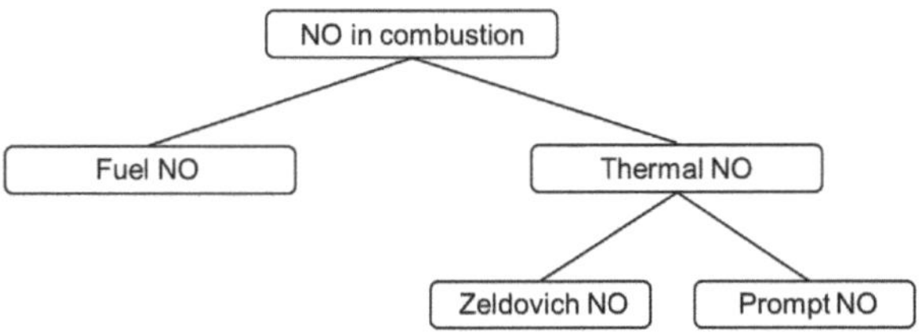

Figure 3.1.1 Classification of NO generated in combustion.

Table 3.1.1 Nitrogen and sulfur in fuel oils.

Type of fuel	N content, mass%	S content, mass%
Raw oil (Middle East)	0.09–0.22	
C heavy oil	0.1–0.4	3.5
A heavy oil	0.05–0.1	0.5
Light oil	0.002–0.03	0.05
Kerosene oil	0.0001–0.0005	0.008
Coal	0.2–3.4	0.2–3
Liquefied petroleum gas, city gas	0	0

"dirty" fuel such as heavy oils and coal. Table 3.1.1 shows nitrogen and sulfur inclusions for various fuels originated from fossil fuels. Note that SO_2 is generated by the reaction between sulfur and oxygen, which is similar to the formation mechanism of fuel NO. The real concentration of fuel NO becomes less than one-third of the 100% conversion value of N present in fuel. For example, for 0.5 mass% nitrogen included in the fuel, the emitted concentration of fuel NO in the exhaust gas becomes 210–420 ppm. Fuel NO is never neglected in heavy oils and coal. As explained previously, the formation of fuel NO greatly depends on oxygen concentrations; fuel NO increases with an increase in oxygen concentration, especially, when air-to-fuel ratio λ becomes large to exceed 1 (excess oxygen state), fuel NO formation rapidly increases. On the other hand, in the excess fuel region of $\lambda < 0$, the formation of fuel NO is suppressed greatly. In low λ combustion, CO, soot (PM), and SO_x tend to increase. When air and fuel are mixed well, e.g., during diffusion combustion in diesel engines, the formation of fuel NO tends to increase compared with premixed combustion.

Thermal NO is generated by the oxidation reaction between N_2 and O_2 in the air at temperatures higher than 1500°C. Thermal NO is classified into Zeldovich NO and prompt NO as shown in Fig. 3.1.1. Zeldovich NO was first identified by Y. B. Zeldovich [4] and is formed outside of the combustion flames. The mechanism of Zeldovich NO formation process is described as follows:

$$O_2 \leftrightarrow O + O \text{ (forward and reverse reaction rates } k_{+1} \text{ and } k_{-1}) \quad (3.18)$$
$$N_2 + O \leftrightarrow NO + N \text{ (forward and reverse reaction rates } k_{+2} \text{ and } k_{-2}) \quad (3.19)$$
$$N + O_2 \leftrightarrow NO + O \text{ (forward and reverse reaction rates } k_{+3} \text{ and } k_{-3}) \quad (3.20)$$

$$N + OH \leftrightarrow NO + H \text{ (forward and reverse reaction rates } k_{+4} \text{ and } k_{-4}) \quad (3.21)$$

Reaction (3.18) represents oxygen radical (O radical) formation at high temperatures. The O radical attacks N_2, then NO and the N radical are formed (reaction (3.19)). The N radical attacks O_2, then NO and the O radical are formed (reaction (3.20)). The process of the chain reactions (3.18)−(3.20) is termed the Zeldovich mechanism. The NO formation process that includes reaction (3.21), i.e., reactions (3.18)−(3.21), is termed the extended Zeldovich mechanism.

Prompt NO, first identified by Fenimore [5], is observed in premixed hydrocarbon flames under the fuel-rich (fuel-excess) condition. The formation of NO is rapid and exceeds the Zeldovich NO formation. Note that the temperature dependency of prompt NO is low. The mechanism of prompt NO formation process is as follows:

$$CH + N_2 \rightarrow HCN + N \quad (3.22)$$

$$N + O_2 \rightarrow NO + O \quad (3.23)$$

$$HCN + OH \rightarrow CN + H_2O \quad (3.24)$$

$$CN + O_2 \rightarrow NO + CO \quad (3.25)$$

In this mechanism, HCN and N radicals are generated as intermediate substances (reaction (3.22)) of the combustion. Prompt NO is finally formed by the oxidation of N and CN in reactions (3.23) and (3.25) inside of the combustion flames.

3.1.4 Calculation of Zeldovich NO formation

Zeldovich NO is formed by combining N_2 and O_2 in the air at high temperatures (> 1500°C) [1] through extended Zeldovich mechanism (reactions (3.18)−(3.21)). The reaction rate in these reactions is the rate-limiting step reaction (3.19) or the slowest one at high temperature. The formation rate of NO by reaction (3.19) is calculated as follows:

$$\gamma_{NO} = 2k_{+2}[N_2][O] \quad (3.26)$$

$$k_{+2} = 13.6 \times 10^{13} \exp\left(-\frac{316.7 \times 10^3}{RT} \right), \left(\frac{cm^3}{mol \cdot s} \right) \quad (3.27)$$

where R is the gas constant (J/K) and T is the absolute temperature (K). The value of k_{+2} can be obtained from the National Institute of Standards and Technology (NIST) Chemical Kinetic Database. The unit of the reaction rate coefficient k_{+2} becomes cm^3/molecule/s or cm^3/mol/s in a two-body first-order reaction. In a three-body first-order reaction, the unit becomes cm^6/molecule/s or cm^6/mol/s. In these units, "/molecule" represents per a single molecule and should be converted to "/mol" using Avogadro's number (=6.022×10^{23} molecule/mol) as follows:

$$1\left(\frac{cm^3}{mol\cdot s}\right) = 1\left(\frac{cm^3}{mol\cdot s}\right) \times \frac{1}{6.022 \times 10^{23}}\left(\frac{mol}{molecule}\right)$$
$$= \frac{1}{6.022 \times 10^{23}}\left(\frac{cm^3}{molecule\cdot s}\right) \quad (3.28)$$

Therefore,

$$k_{+2} = 2.26 \times 10^{-10} \exp\left(-\frac{316.7 \times 10^3}{RT}\right), \left(\frac{cm^3}{molecule\cdot s}\right) \quad (3.29)$$

It is noted that in some works of literature, the units of cm^3/molecule/s and cm^3/mol/s are simply written as cm^3/s. Because O_2 and O are at an equilibrium state downstream of the combustion flame, the following relations are materialized with an equilibrium constant K

$$k_{+1}[O_2] = k_{-1}[O]^2, \ \frac{[O]}{[O_2]^{\frac{1}{2}}} = \left(\frac{k_{+1}}{k_{-1}}\right)^{\frac{1}{2}} = K, \ \therefore [O] = K[O_2]^{\frac{1}{2}} \quad (3.30)$$

where

$$K = 5.06 \exp\left(-\frac{248 \times 10^3}{RT}\right), \left(\frac{mol^{\frac{1}{2}}}{cm^{\frac{3}{2}}}\right) \quad (3.31)$$

Combining Eqs. (3.26), (3.27), (3.30), and (3.31), the formation rate of NO is calculated using Eq. (3.32).

$$\gamma_{NO} = k'[N_2][O_2]^{\frac{1}{2}}, \left(\frac{mol}{cm^3\cdot s}\right) \quad (3.32)$$

where

$$k' = K \times 2k_{+2} = 137.6 \times 10^{13}\exp\left(-\frac{564.7 \times 10^3}{RT}\right), \left(\frac{cm^{\frac{3}{2}}}{mol^{\frac{1}{2}}\cdot s}\right) \quad (3.33)$$

Example calculation of Zeldovich NO: Using Eqs. (3.32) and (3.33), the formation rate of Zeldovich NO, γ_{NO}, (unit: ppm/s) at a temperature T (°C) is derived. In this example calculation, Celsius temperature is expressed with T. Care must be taken while converting the unit. For example in Eq. (3.32), the units of $[N_2]$ and $[O_2]$ are mol/cm^3. The molar fraction in the atmosphere is approximately $[N_2] = 79\%$ and $[O_2] = 21\%$ that must be converted to the unit of mol/cm^3. It is noted that at temperature T, 1 mol is equivalent to $22.4 \times 10^3 \times (T+273.15)/273.15$ cm^3; therefore,

$$[N_2] = \frac{0.79}{22.4 \times 10^3 \times (T + 273.15)/273.15}, \left(\frac{\text{mol}}{\text{cm}^3}\right) \tag{3.34}$$

$$[O_2] = \frac{0.21}{22.4 \times 10^3 \times (T + 273.15)/273.15}, \left(\frac{\text{mol}}{\text{cm}^3}\right) \tag{3.35}$$

Because the unit of γ_{NO} is mol/cm^3/s in Eq. (3.32), it should be converted to the unit of ppm/s as follows:

$$\begin{aligned}\gamma_{NO}\left(\frac{\text{ppm}}{\text{s}}\right) &= 137.6 \times 10^{13} \exp(-564.7 \times 10^3/8.314/[T + 273.15]) \\ &\times \left[\frac{0.79}{22.4 \times 10^3 \times (T + 273.15)/273.15}\right] \\ &\times \left[\frac{0.21}{22.4 \times 10^3 \times (T + 273.15)/273.15}\right]^{\frac{1}{2}} \\ &\times 22.4 \times 10^3 \times \frac{T + 273.15}{273.15} \times 10^6 \\ &= \frac{5.501 \times 10^{19}}{(T + 273.15)^{\frac{1}{2}}} \exp(-6.792 \times 10^4/[T + 273.15])\end{aligned} \tag{3.36}$$

The relation between the NO formation rate and Celsius temperature T based on Eq. (3.36) is shown in Fig. 3.1.2. Zeldovich NO significantly increases at temperatures greater than 1500°C and the ratio $\gamma_{NO}(1600°C)/\gamma_{NO}(1500°C)$ becomes 7.52.

3.1.5 Other nitrogen oxides

NO_x represents mainly nitric monoxide (NO) and nitrogen dioxide (NO_2), but other nitrogen oxides exist such as dinitrogen dioxide (N_2O_2), dinitrogen trioxide (N_2O_3), dinitrogen tetroxide (N_2O_4), and dinitrogen

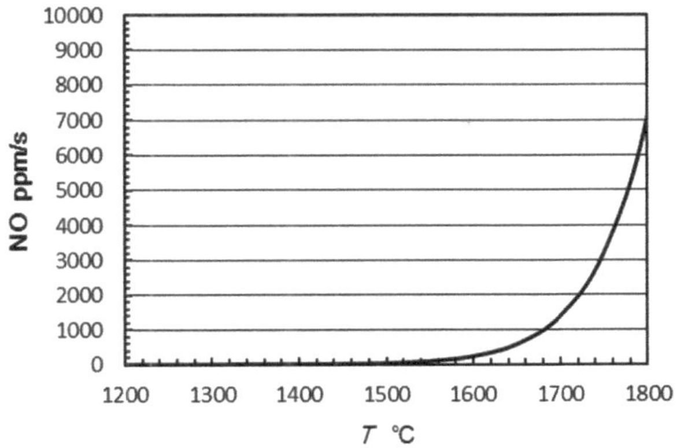

Figure 3.1.2 Relation between NO formation rate and temperature.

pentoxide (N_2O_5). At the exit of the exhaust pipe of diesel engines, 90–95% of NO_x is NO. When it is emitted to the atmosphere, it is oxidized to NO_2 and N_2O_5. They are easily dissolved in water and rain to induce acid rain; therefore, they should be strictly controlled. N_2O (nitrous oxide) is usually not included in NO_x, but it is one of the greenhouse gases with a global warming potential (GWP) approximately 300 times that of carbon dioxide. N_2O is also a stratospheric ozone-depleting gas. Approximately 80% of N_2O emissions are associated with microbial activity on land and water, and approximately 12% is also generated during low-temperature (700–900°C) combustion of coal and oil.

Two materials contributed to N_2O production during the combustion are known. One is a contribution of NH_3 in the reaction processes (3.37)–(3.39).

$$NH_3 + OH \rightarrow NH_2 + H_2O \tag{3.37}$$
$$NH_2 + OH \rightarrow NH + H_2O \tag{3.38}$$
$$NH + NO \rightarrow N_2O + H \tag{3.39}$$

Because urea or NH_3 is injected into the combustion gas in the selective noncatalytic reduction (SNCR) and selective catalytic reduction (SCR) aftertreatment processes, which are explained later, there is a tendency of increasing N_2O in the exhaust gas. The other is a contribution of HCN in the combustion flame and gas as shown in reactions (3.40)–(3.42).

$$HCN + OH \rightarrow HNCO + H \tag{3.40}$$
$$HNCO + OH \rightarrow NCO + H_2O \tag{3.41}$$
$$NCO + NO \rightarrow N_2O + CO \tag{3.42}$$

At temperatures > 1500 K, N_2O is decomposed by H and OH radicals in the combustion process

$$N_2O + H \rightarrow N_2 + OH \tag{3.43}$$

$$N_2O + OH \rightarrow N_2 + HO_2 \tag{3.44}$$

To suppress the formation of N_2O, the combustion temperature should be kept high (> 1500 K), and in the SCR aftertreatment, urea or NH_3 injection at the temperature of 1000 K should be avoided. Future studies are needed to reduce N_2O emissions that should be regulated from the viewpoint of global environmental protection.

3.2 NO_x control

Decomposition of nitrogen oxide (De-NO_x) aftertreatment is generally classified into two groups, wet process and dry process. In the wet process, which generally uses a chemical scrubber, NO is first oxidized by ozone injection or oxidation catalysts to NO_2 or N_2O_5 and finally converted to HNO_3 or reduce to N_2 by reducing agents such as Na_2SO_3. When the temperature at the exit of the chemical scrubber is higher than the dew point temperature or more than 200°C, the process is called semidry because the gas becomes dry there. On the other hand, in the dry processes using catalysts and/or adsorbents, the following methods are known: (1) NH_3 or hydrocarbon (HC) is injected into the exhaust gas to reduce NO_x at the gas phase or on solid catalysts and (2) activated carbon or molecular sieve (zeolite) is used to adsorb NO_x. Although the wet process was vigorously investigated at the initial stage of the development of the De-NO_x technology for combustors, the dry processes using catalysts now become popular, especially for large combustors. However, the wet and semidry processes are convenient for relatively small combustors, the flue gas flow rate of which is less than 10,000 Nm^3/h although several drawbacks exist with its application.

Because NO has low water solubility, it should be oxidized to easily water-soluble NO_2 or N_2O_5 either by the oxidation catalysts such as platinum (Pt) or palladium (Pd) or ozone injection generated by the NTP process. In coal- or heavy oil-fired combustors, where SO_x in the exhaust gas damages the catalysts (catalyst poisoning), the ozone injection method is very useful for the oxidation of NO. However, a relatively large amount of energy is required for ozone formation in the plasma reactor. Moreover, a large adsorption tower is required in the secondary wet process, and a large amount of exhaust gas should be absorbed by the solution. Furthermore, at the end of the wet chemical process, water solution treatment is necessary because alkali or acid solution should be processed according to the regulation before the solution is drained. However, these drawbacks can be resolved recently with the developments of ozone generator energy efficiency and the semidry process as explained in Subsection 5.1.2 to the application for the exhaust gas treatment of glass melting furnace.

In contrast, in dry processes such as SCR using NH_3 reducing agents, because the chemical reaction rate is faster than that of the wet process, a smaller reactor can treat the exhaust gas. Furthermore, waste solution treatment is not necessary because NO is converted to harmless N_2 under the dry state. These advantages made the dry processes the main treatment technology for combustors on land. However, for marine diesel engines aftertreatment, there is a possibility that both wet and dry processes are used because of plenty of seawater. Further, lack of reducing agents such as urea is a problem during ship moving. Therefore, the technology is under development for both dry and wet methods.

3.2.1 Selective noncatalytic reduction

NO can be reduced to N_2 by reducing agents such as NH_3, HC, CO, and H_2. However, under oxygen-rich exhaust conditions ($O_2 > 1\%$) such as in combustion process, NO is mainly oxidized to NO_2, NO_3, and N_2O_5, and scarcely reduced to N_2. In the 1970s, Lyon [6] found that under oxygen-rich exhaust conditions, at temperatures 800–1000°C, NH_3 can reduce NO selectively. This finding has greatly contributed to the development of the reduction technologies of NO_x in the flue gas. In this subsection, this thermal De-NO_x aftertreatment process called SNCR is explained. NO reduction reaction by NH_3 is described as

$$4NO + 4NH_3 + O_2 \rightarrow 4N_2 + 6H_2O \tag{3.45}$$

In this reaction, O_2 is necessary for the chemical reactions of NO. At the temperature range of 800–1000°C, which is a narrow window-type range, highest NO_x removal is achieved. In recent research, OH and H radicals have played an important role in NO_x reduction. SNCR technology for NO_x is an application of these mechanisms. First, urea or NH_3 is injected into the exhaust pipe where the gas temperature is 900–1000°C under oxygen-rich conditions. As a result, NO is reduced to N_2. The efficiency of reduction is typically 90% and the reaction time is approximately 0.5 s.

However, in this method, the temperature control at the injection point which often located inside the combustor is difficult and sometimes NO_x concentrations may increase by the thermal decomposition of NH_3. In recent years, SNCR has been used for exhaust gas De-NO_x of garbage incinerators and glass melting furnaces. Ammonia or urea is sprayed to the furnace in an appropriate temperature range, giving successful results. Furthermore, this method is difficult to apply for the control of emissions from marine diesel engines because the exhaust gas temperature of diesel engines is typically 300−400°C, which is a lower temperature range for SNCR.

3.2.2 Selective catalytic reduction

The application of SNCR is restricted by the high-temperature range requirement. In the SCR process, NO is reduced selectively by solid SCR catalysts using ammonia (NH_3) at lower temperatures, typically 300°C. The mechanism of SCR process on the typical catalyst of vanadium pentoxide (V_2O_5) is shown in Fig. 3.2.1. The figure schematically shows a cyclical NO reduction process. As shown in Fig. 3.2.1A, vanadium pentoxide catalyst has O or OH on the surface, and NO is removed using NH_3. As shown in Fig. 3.2.1B, vanadium pentoxide is once activated at O and OH on the surface. Among the connections in the figure, the broken line means

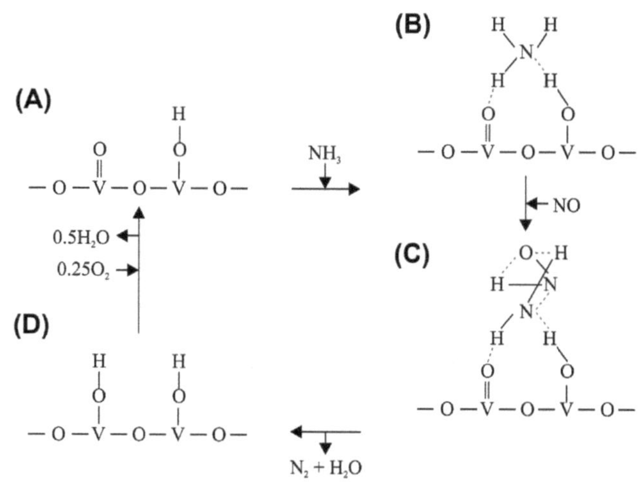

Figure 3.2.1 Reaction mechanism of selective catalytic reduction of NO with ammonia [7]. (A) Vanadium pentoxide catalyst with O and OH on the surface, (B) ammonia reacting with the catalyst surface, (C) NO selectively reacting with NH_3, and (D) vanadium pentoxide catalyst with two OH on the surface. Dotted connections mean activated ones.

that interaction is occurring. Ammonia reacts with the catalyst surface. As shown in Fig. 3.2.1C, N_2 and H_2O bonds are formed on the catalyst by the reaction of activated ammonia with NO. NO can react with ammonia although the high concentration O_2 in the atmosphere cannot react with it. Therefore, NO selectively reacts with NH_3 by this unique catalyst system. As shown in Fig. 3.2.1D, consequently the final products of N_2 and H_2O are discharged. Although two OH are left on the surface of the catalyst, one is reduced to O by discharging H_2O and O_2. The reactions are cyclically repeated. This cycle utilizes the reduction–oxidation (redox) reaction of vanadium pentoxide catalyst. The catalyst is often loaded on TiO_2 because it can prevent SO_x remaining in the exhaust gas from becoming sulfate.

A schematic diagram of SCR for the combustor is shown in Fig. 3.2.2. Ammonia is injected before the gas is passed through the catalyst. SCR is followed by the flue gas desulfurization system to remove SO_x in the exhaust gas. Cleaned exhaust gas is emitted to the atmosphere at the stack.

The exhaust gas mixed with NH_3 flows into the SCR catalyst having a honeycomb structure. On the surface of the catalyst, SCR of NO_x is carried out by NH_3. During the process, the temperature of the reactor should be maintained at 200–400°C. For catalysis, either V_2O_5 or CuO is typically loaded on TiO_2 ceramics. Usually, TiO_2 is used because of its higher durability against sulfur. In this method, NO removal efficiency often exceeds 90% when the space velocity (SV) equals to $10^4\ h^{-1}$, where SV is defined as follows:

$$SV = \frac{Q}{V}\left(\frac{m^3/h}{m^3}\right) = \frac{1}{T_r(h)} \tag{3.46}$$

where Q is the flow rate of the exhaust gas, V is the volume of the catalyst, and T_r is the residence time. Therefore, SV is defined as the volume of the exhaust gas for 1 m^3 catalyst per hour. In the SCR method, dustproof catalysts should be used for marine diesel engines fueled with heavy oil. The SCR method is very popular for the treatment of thermal electric power stations and recently for marine diesel engines because NO_x emission

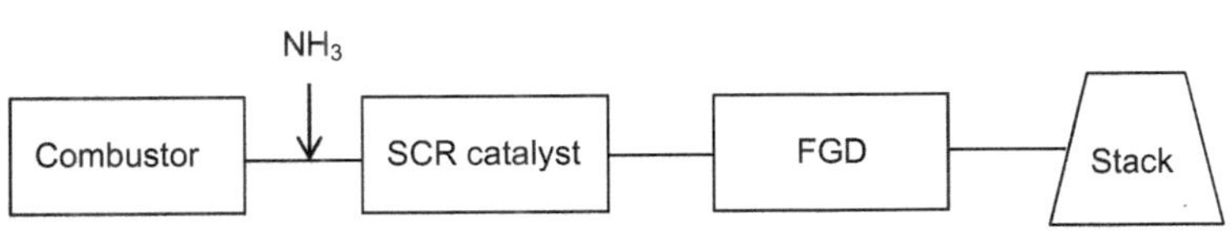

Figure 3.2.2 Total emission control system for combustors. *SCR*, selective catalytic reduction; *FGD*, flue gas desulfurization.

regulations have become very stringent since 2016. Because residual NH_3 after NO reduction induces corrosion with acidic sodium sulfate, too much NH_3 or urea injection should be avoided.

3.2.3 Urea selective catalytic reduction for marine diesel engines

Diesel engines mounted in large ships, such as freighters and tankers, are of two types: the main propulsion unit used to drive the ship (two-cycle engines in most cases) and the auxiliary generator used to supply electrical power inside the ship (four-cycle engines in most cases). Generally, there are several engines of these types inside the ship. Although exhaust gases can flow together through a single stack, each engine should be fitted with its own exhaust gas treatment device. For NO_x reduction, the urea SCR method is employed for marine diesel engines because urea injection is safer than ammonia injection. Ammonia is dangerous gas and should be avoided in the ship loading.

Fig. 3.2.3 shows a schematic of SCR for De-NO_x of marine diesel engine exhaust gas. A honeycomb SCR catalyst (typically V_2O_5 supported on a TiO_2 base material) is placed inside the path of the exhaust gas from the diesel engine. An aqueous urea solution stored in the urea tank is sprayed from the urea solution nozzles placed several tens of centimeters upstream (process (a) in the figure). The aqueous urea solution is hydrolyzed into ammonia and carbon dioxides according to the following reaction of urea solution.

$$CO(NH_2)_2 + H_2O \rightarrow CO_2 + 2NH_3 \tag{3.47}$$

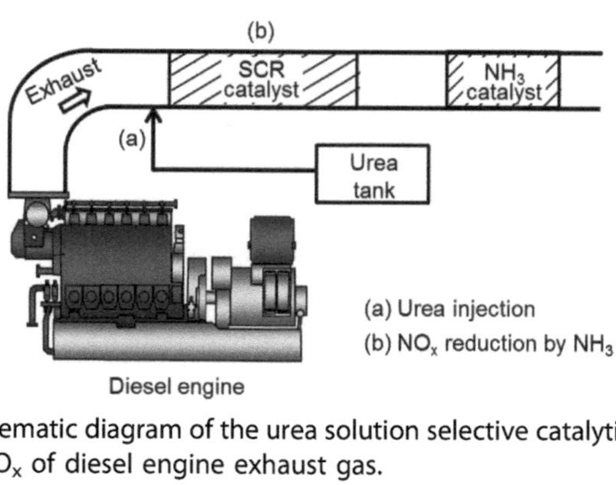

Figure 3.2.3 Schematic diagram of the urea solution selective catalytic reduction (SCR) system for De-NO_x of diesel engine exhaust gas.

Urea aqueous solution of 33 mass% is commercialized. NO_x is cleaned via reaction with NH_3 and reduced to N_2 (reactions (3.48) and (3.49), process (b) in the figure).

$$4NO + 4NH_3 + O_2 \rightarrow 4N_2 + 6H_2O \tag{3.48}$$

$$6NO_2 + 8NH_3 \rightarrow 7N_2 + 12H_2O \tag{3.49}$$

Unreacted NH_3 left in the exhaust conduit is referred to as the NH_3 slip, and its emissions must be suppressed by decomposition using an NH_3 catalyst for oxidation, placed downstream of the SCR catalyst. As an example of a performance verification test, when the exhaust gas temperature is 350°C and the SV is of 10,000 h^{-1}, the NO_x removal efficiency more than 80% can be obtained.

3.2.4 Selective noncatalytic reduction assisted by nonthermal plasma

The SNCR technology is a simple and cost-effective process that involves the injection of molecular ammonia without the use of catalysts. In a conventional SNCR system, it is known that a NO_x reduction occurs at temperatures of 850–1175°C, whereas the SCR is effective at temperatures of 250–400°C. This temperature range is called the "temperature window." However, sufficiently high efficiency of the NO_x reduction cannot be obtained in large-scale combustors because the range is too high and narrow to realize a De-NO_x reaction. To improve the NO_x reduction efficiency of SNCR, expanding the temperature window is necessary to incorporate lower temperatures. Assistance of NTP induced by a dielectric barrier discharge (DBD) is an alternative technique used to solve this problem. The SNCR of NO_x by activated ammonia generated using an intermittent DBD has been developed by Kambara et al. [8] to broaden and lower the narrow temperature window for De-NO_x.

Fig. 3.2.4 shows a schematic of the De-NO_x test facility for SNCR assisted by DBD-induced plasma. In the figure, the length is shown in millimeters such as $L = 360$ and $L = 490$. In addition, HV stands for high voltage, and GC, PAS, and MFC indicate gas chromatography, photoacoustic spectroscopy, and mass flow controller, respectively. The apparatus consists of two heater furnaces with quartz tubes, gas mixing and flow control systems, a plasma reactor, a high-voltage pulse power supply, and gas analyzers. The heater furnaces with quartz tubes are named as a preheater quartz tube and a reaction quartz tube with lengths of 500 and

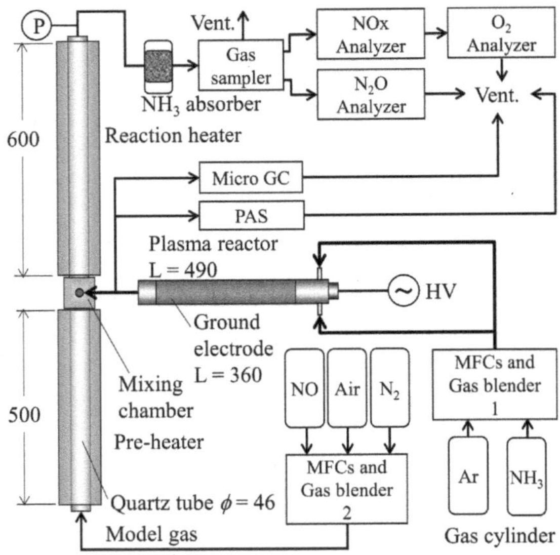

Figure 3.2.4 Schematic of experimental apparatus of selective noncatalytic reduction assisted by dielectric barrier discharge–induced nonthermal plasma [8].

600 mm, respectively. The quartz tubes are connected with the mixing chamber, which features a heat insulator. A $NO/O_2/N_2$ gas mixture is prepared using MFCs and a gas blender and is fed into the preheater quartz tube as the model flue gas. The pressure of the reaction quartz tube is controlled at slightly above atmospheric pressure using a gas sampler. The electrodes of the plasma reactor are coaxial in configuration, with quartz tubes used as the dielectric materials. The outer quartz tube is 61 mm in diameter, 2 mm in thickness, and $L = 490$ mm in length, whereas the inner quartz tube is 54 mm in diameter, 2 mm in thickness, and $L = 490$ mm in length. NTP is generated in a 1.5 mm gap. A high-voltage electrode, 50 mm in diameter and $L = 490$ mm in length, is positioned inside the inner quartz tube. A grounded electrode, $L = 360$ mm in length and 0.2 mm in thickness, covers the outer surface of the outer quartz tube. Ammonia gas diluted with argon is supplied to the gap of the plasma reactor. Activated ammonia, generated by the DBD-induced NTP, is introduced into the mixing chamber through a polytetrafluoroethylene (PTFE) pipe and streamed into the reaction quartz tube along with the model flue gas.

As a result, a temperature window enlargement of 150°C is achieved at the lower boundary of the window using an activated ammonia injection, that is, the window reaches 700–1175°C. The reaction mechanism for De-NO_x is considered based on measurements of the chemical composition of activated ammonia, De-NO_x verification experiments, and a detailed kinetics simulation. The results indicate that a major chemical species of activated ammonia is molecular hydrogen. The effects of molecular hydrogen on De-NO_x reactions are investigated using an NH_3/H_2 gas mixture within the temperature range of 500–850°C, which overlaps the enlarged temperature window. The result clearly indicates that hydrogen plays a key role in the expansion of the temperature window. The reaction mechanism of De-NO_x can be described using a detailed elemental kinetic model. This model also suggests that NH_3 decomposition takes place firstly, and H_2 is formed in plasma. Next, molecular hydrogen produces a key radical species for De-NO_x reactions, such as HNO, NNH, or NH, at above 600°C, which promotes a NO_x reduction at the lower temperatures (= 600°C) of the enlarged temperature window. To apply this method to the exhaust gas treatment of diesel engines, the typical temperature of which is 400°C, it is necessary to further lower the operating temperature. Efforts are continuing.

3.2.5 Diesel particulate-NO_x reduction for automobile emission

Diesel particulate-NO_x reduction (DPNR) for automobile diesel engines is based on NO_x storage and reduction (NSR) or lean NO_x trap for NO_x reduction, but PM is also reduced simultaneously. In the following, the NSR process is firstly explained, and then PM and NO_x simultaneous reduction together with the NSR, namely, the DPNR process, is explained.

NO_x storage and reduction process

The NSR catalyst plays an important role in NO_x reduction. The NSR technology has been in development since 1995 [9]. It works under the cyclic operations of fuel-lean and fuel-rich conditions. Fuel-lean state is realized during normal engine operation, while fuel-rich state is realized by periodic fuel injection in the exhaust gas. The NSR catalyst consists of noble metals, NO_x storage materials, and support metal oxides. In general, noble metals and NO_x storage materials are dispersed on support metal oxides. The metal oxides such as Al_2O_3, ZrO_2, CeO_2, MgO, or mixed metal oxides such as $MgO-CeO_2$ and $MgO-Al_2O_3$ are used as the support metal oxides [10]. Although several NSR catalysts have been proposed and developed, the most popular one is $Pt/BaO/Al_2O_3$. NO_x reduction in the NSR catalyst is achieved by the following two processes: Process 1 is

NO oxidation to NO_2 and storage under fuel-lean conditions (air- and O_2-rich conditions) during the normal fuel-lean engine operation. Process 2 is NO_x reduction to N_2 under fuel-rich conditions (air- and O_2-lean conditions) during the periodic fuel-rich engine operation. Fig. 3.2.5 depicts process diagram of NSR over NSR catalyst. The details of processes 1 and 2 are explained next.

Process 1: In general, NO_x storage materials work more effectively for NO_2 than NO. However, typically 90% of NO_x exists as NO in the diesel exhaust gas. Therefore, the pretreatment for the oxidation of NO to NO_2 is an important process. Under fuel-lean conditions in normal engine operation where air excess ratio is $\lambda > 1$, NO in the exhaust gas is initially oxidized into NO_2 on the surface of noble metals:

$$NO + O_2 \rightarrow NO_2 + O \tag{3.50}$$

This oxidation reaction proceeds via a four-step mechanism with a noble metal [9]:

$$\text{Step 1: } NO + Pt \leftrightarrow NOPt \tag{3.51}$$

$$\text{Step 2: } O_2 + 2\ Pt \leftrightarrow 2O{-}Pt \tag{3.52}$$

$$\text{Step 3: } NO{-}Pt + O{-}Pt \leftrightarrow NO_2{-}Pt + Pt \tag{3.53}$$

$$\text{Step 4: } NO_2{-}Pt \leftrightarrow NO_2 + Pt \tag{3.54}$$

It is noted that NO−Pt, O−Pt, and NO_2−Pt means that NO, O, and NO_2 are adsorbed on Pt. The O radical is dissociated from the O−Pt in step 3. Noble metals such as platinum (Pt), palladium (Pd), and ruthenium (Rh)

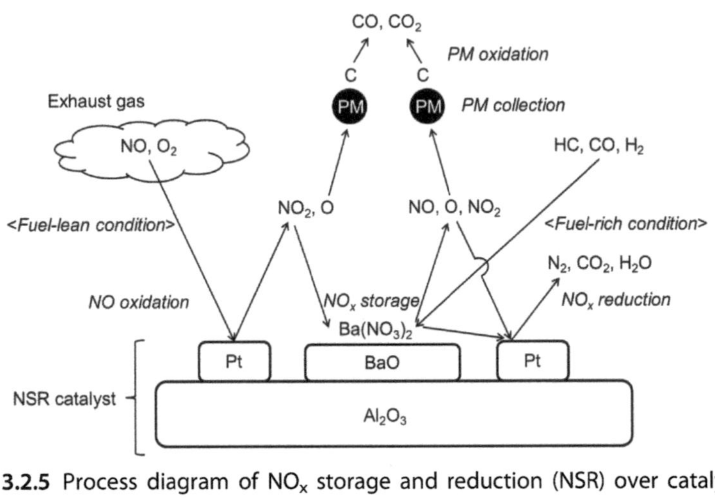

Figure 3.2.5 Process diagram of NO_x storage and reduction (NSR) over catalyst and diesel particulate reduction.

are used in low amounts of 1–2% by mass. These noble metals play two significant roles in the NSR mechanism, one is the oxidation of NO under fuel-lean conditions and the other is the reduction of the stored NO_x under fuel-rich conditions. Pt shows higher performance for NO oxidation, whereas Pd or Rh shows higher performance for NO_x reduction. In view of these characteristics, bimetallic NSR catalysts such as Pd/Rh and Pt/Rh have also been proposed and used [10]. In NO oxidation, experimental data of the steady-state and equilibrium NO–NO_2 conversions are provided in the temperature range of 90–500°C on the Pt/BaO/Al_2O_3 catalyst [9]. On the condition that the actual inlet NO and O_2 volume concentrations are 528 ppm and 5%, the actual NO conversion efficiency agrees with the equilibrium NO conversion one at higher temperatures than 350°C. On the condition, the actual NO conversion efficiency is 60% at 350°C. At higher temperatures (> 350°C), for instance, the actual and equilibrium NO conversion efficiencies are typically 40% and 20% at 430 and 500°C, respectively.

Induced NO_2 is temporally stored in NO_x storage materials, which are generally alkali metals or alkaline earth metal compounds. Among them, barium (Ba)-based storage materials are mostly used. Ba loading typically varies from 8% to 20%. Other alkali metals such as Na and K or alkaline earth metals such as Mg, Sr, and Ca are also used. The basicity of the alkali or alkaline earth metal components is directly related to the performance of NO_x storage. The performance decreases in the order of K > Ba > Sr ≥ Na > Ca > Li ≥ Mg at 350°C [10]. The NO_x storage process is a three-step mechanism in the case of barium oxide (BaO) use. At the first step, the oxidized NO_2 is adsorbed slowly on BaO and becomes BaO–NO_2. At the second step, BaO–NO_2 decomposes into barium peroxide (BaO_2) and NO. NO is released to the gas phase. At the third step, BaO_2 reacts with the oxidized NO_2 and becomes barium nitrate ($Ba(NO_3)_2$). This NO_x storage process is given by the following chemical reactions:

$$BaO + NO_2 \rightarrow BaO{-}NO_2 \tag{3.55}$$

$$BaO{-}NO_2 \rightarrow BaO_2 + NO \tag{3.56}$$

$$BaO_2 + 2NO_2 \rightarrow Ba(NO_3)_2 \tag{3.57}$$

Process 2: Under fuel-rich conditions, which is realized by periodic fuel injections in the exhaust gas, when reduction components such as hydrocarbons (HC, e.g., C_3H_6 or C_3H_8), CO, and H_2 are regularly introduced; the stored NO_x is released, and then it reacts with HC and CO to be converted into N_2, CO_2, and H_2O. This is the reduction process for converting NO_x into N_2. For instance, fuel-rich conditions are realized by the controlled common rail fuel injection. NO_x reduction takes place on the Pt surface in a very short time (a few seconds). The performance depends on the type and amount of reduction components and operating temperature. Depending on the conditions, the reduction process could generate not only NO but also NO_2, N_2O, and NH_3 as byproducts. The byproduct of NO_2 is observed at a very short regeneration time. N_2O and NH_3 are observed at low temperatures of 200–300°C and long regeneration times.

Reducing gases such as CO or HCs can reduce the stored NO_x either directly to form N_2 in the storage component or indirectly by forming H_2 via the water–gas shift reaction ($CO + H_2O \rightarrow CO_2 + H_2$) to form N_2 [10]. HCs can produce H_2 as well. This process is called the steam reforming reaction. Under fuel-rich conditions, H_2 itself exists in the exhaust gas. The direct reduction of NO_x, which is stored as $Ba(NO_3)_2$, takes place as

$$Ba(NO_3)_2 + 5CO \rightarrow N_2 + BaO + 5CO_2 \tag{3.58}$$

The indirect reduction takes place via H_2 from the water–gas shift or the steam reforming reactions as

$$Ba(NO_3)_2 + 5H_2 \rightarrow N_2 + BaO + 5H_2O \tag{3.59}$$

NO_x reduction performance is the highest for H_2, followed by CO, C_3H_6, and C_3H_8. Depending on the catalyst, nonreduced NO, N_2O, and NH_3 can also be formed as byproducts as

$$Ba(NO_3)_2 + 3H_2 \rightarrow 2NO + BaO + 3H_2O \tag{3.60}$$

$$Ba(NO_3)_2 + 4H_2 \rightarrow N_2O + BaO + 4H_2O \tag{3.61}$$

$$Ba(NO_3)_2 + 8H_2 \rightarrow 2NH_3 + BaO + 5H_2O \tag{3.62}$$

These processes of stored NO_x reduction by H_2 take place directly as reaction (3.59) and indirectly via various paths (reaction network) with intermediate species such as NH_3 and N_2O [11].

Sulfur degradation or poisoning of NSR catalysts is a significant issue in the reduction process. SO_2 in the exhaust gas is oxidized on noble metals and reacts with the support metal oxides. As a result, aluminum sulfate ($Al_2(SO_4)_3$) is formed. As another possibility, SO_x reacts with the NO_x storage component, and barium sulfate ($BaSO_4$) is formed. BaO can form

$BaSO_4$ in oxidation reactions and $BaSO_3$ in reduction reactions. These sulfates are more stable than nitrates. Therefore, they depress the performance of NO_x adsorption. This is the cause of the deactivation of the catalyst.

Diesel particulate–NO_x reduction process

In diesel emissions, it is difficult to reduce both PM and NO_x simultaneously by improving the performance of combustion. The relationship between PM and NO_x in emissions is a trade-off and a serious issue for diesel engines. The DPNR method is introduced as one of the aftertreatment technologies to solve this problem. Up to date, the technology has been applied to automobiles. A DPNR catalyst, which is a ceramic diesel particulate filter (DPF) coated with NSR catalysts, is key to reducing both PM and NO_x. Fig. 3.2.5 shows the complete process diagram of diesel particulate and NO_x reduction.

Soot PM (diesel particulates) that consists of carbon (C) is firstly collected by a DPNR catalyst with DPF. The soot PM is oxidized by O radical and NO_2 at temperatures higher than 250°C. This process is the regeneration of the DPNR catalyst. The oxidation reactions of soot PM with O radical are given by reactions (3.63) and (3.64):

$$C + 2O \rightarrow CO_2 \tag{3.63}$$

$$C + O \rightarrow CO \tag{3.64}$$

These oxidation reactions proceed when the NO oxidation reaction (3.50) and reduction processes proceed on the NSR catalyst. The oxidation reactions of soot PM with NO_2 and NO under fuel-lean conditions are presented by

$$C + NO_2 \rightarrow CO + NO \tag{3.65}$$

$$C + 2NO_2 \rightarrow CO_2 + 2NO \tag{3.66}$$

$$C + NO_2 + O_2/2 \rightarrow CO_2 + NO \tag{3.67}$$

The oxidized NO_2 is reduced to NO in the presence of soot PM. This is a complete description of DPNR method.

Plasma-enhanced DPNR process

For the simultaneous reduction of NO_x and oxidation of soot PM under fuel-rich conditions, where O_2 concentration is less than 10%, the following reactions are desirable [12].

$$C + NO_2/2 \rightarrow CO + N_2/4 \tag{3.68}$$

$$C + NO_2 \rightarrow CO_2 + N_2/2 \tag{3.69}$$

These chemical reactions rarely occur under fuel-lean (oxygen-rich) conditions in diesel emissions. To enhance the simultaneous reduction of NO_x and oxidation of soot PM, an oxygen-lean NTP method that combines DPNR and NTP has been proposed [12]. Under fuel-lean conditions, NO_x reduction is performed on NSR catalyst and soot PM is oxidized similar to the typical DPNR method. When a constant amount of NO_x is stored in the NO_x storage material, fuel-rich conditions are realized by decreasing the ratio of air-to-fuel and/or the exhaust gas recirculation (EGR). Under fuel-rich conditions, NTP is applied to the NO_x storage material for releasing the stored NO_x as NO_2. The released NO_2 is reduced to N_2 by the plasma application. Simultaneously, the collected soot PM and HCs are oxidized by the O radical into CO_2 and H_2O, respectively.

Up to date, the DPNR method has been applied to automobiles, but not to marine diesel engines. There is a problem that these catalyst systems are weak for the sulfur in the exhaust gas. The catalyst, especially the oxidation catalyst, is easily degraded by the sulfur. It is best to use ultra-low-sulfur diesel (ULSD) as fuel for the engine. However, the DPNR method is one of the promising technologies to reduce both PM and NO_x, simultaneously. The DPNR method is a viable technique to be applied to marine diesel engines in the future as the regulations become more stringent and are extended to soot PM.

3.2.6 Exhaust gas recirculation

NO_x reduction mechanism in exhaust gas recirculation

Exhaust gas recirculation (EGR) is one of the emission control technologies for reducing NO_x emissions by recirculating a part of the exhaust gas, and reducing fuel consumption and pumping loss in the meantime. In an engine, the pressure of exhaust gas is greater than that of the intake air. In other words, an engine works as a pump and consumes energy. This energy consumption is recognized as a pumping loss. In the EGR, exhaust gas emitted from diesel engines is recycled and cooled down to introduce the combustion air at the inlet of diesel engines. As a result, combustion temperature may decrease to reduce Zeldovich NO. Prompt NO is also affected by O_2 concentration in combustion. The EGR method is suitable for diesel engines with large exhaust gas flow rates. O_2 contributes to the formation of prompt NO and Zeldovich NO. A/F or AFR is the mass ratio of air to fuel and expressed as

$$A/F = \frac{m_{\text{air}}}{m_{\text{fuel}}} \tag{3.70}$$

where m_{air} and m_{fuel} are the mass flow rates of air and fuel, respectively. The theoretical stoichiometric air-to-fuel ratio $(A/F)_{\text{stoich}}$ is 14.7–14.9 for gasoline and diesel engines. At higher O_2 concentrations such as fuel-lean combustion or lean-burn, the combustion temperature becomes higher, and the amount of thermal NO in the emissions increases. Air excess ratio, λ, is the ratio of the mass m_{air} of actual air to the mass $(m_{\text{air}})_{\text{stoich}}$ of stoichiometric air in combustion, and it is defined as

$$\lambda = \frac{m_{\text{air}}}{(m_{\text{air}})_{\text{stoich}}} = \frac{A/F}{(A/F)_{\text{stoich}}} \tag{3.71}$$

$\lambda > 1.0$ represents air excess (fuel-lean), $\lambda = 1.0$ stoichiometric, and $\lambda < 1.0$ fuel-excess conditions for combustion. It is noted that the right equation of Eq. (3.71) persists when m_{fuel} is constant. The equivalence ratio is the ratio of the fuel-to-oxidizer ratio to the stoichiometric fuel-to-oxidizer ratio.

$$\phi = \frac{m_{\text{fuel}}/m_{\text{oxidizer}}}{(m_{\text{fuel}}/m_{\text{oxidizer}})_{\text{stoich}}} \tag{3.72}$$

where m is the mass flow rate. $\phi > 1.0$ represents fuel-excess, $\phi = 1.0$ stoichiometric, and $\phi < 1.0$ fuel-lean conditions for combustion. In the case where the air is used as the oxidizer, ϕ equals to $1/\lambda$ deriving form Eqs. (3.70)−(3.72) as follows:

$$\phi = \frac{m_{\text{fuel}}/m_{\text{air}}}{(m_{\text{fuel}}/m_{\text{air}})_{\text{stoich}}} = \frac{(A/F)_{\text{stoich}}}{A/F} = \frac{1}{\lambda} \tag{3.73}$$

Diesel engine is considered lean-burn when the air excess ratio, λ, is in the range of 1.5−1.8 on full load [13]. Air excess ratio, λ, becomes higher as the load decreases. At idle conditions, the air excess ratio becomes $\lambda > 10$.

In the thermal NO_x formation mechanism, if the concentration of O_2 in the intake air is reduced or in the rich-burn state, the combustion temperature and NO emission decrease. In addition, because, in EGR, a part of the exhaust gas is introduced into the engine intake, CO_2 and H_2O in the exhaust gas are also introduced. CO_2 and H_2O, which have a higher molar heat capacity than N_2 and O_2, can contribute to the reduction of the combustion temperature. Although the EGR method is very effective for the reduction of NO, the following problems may exist. (1) It is difficult to reduce fuel NO because the effect of temperature on fuel NO is small. (2) The combustion process becomes unstable. (3) The cost for EGR apparatus is high.

Implementation of exhaust gas recirculation

The fundamental principle of EGR is to recycle a part of exhaust gas into the intake of an engine. If the exhaust gas is directly recirculated into the engine intake, the operation is called hot EGR. In the hot EGR, the density of the hot exhaust gas is lower than the fresh intake air. If the exhaust gas is recirculated into the engine intake after being cooled by the EGR cooler, the operation is called cooled EGR. On the other hand, EGR processes are classified into internal and external EGRs. They are explained next.

The procedure of the internal EGR is shown in Fig. 3.2.6. The internal EGR utilizes a valve overlap in a four-cycle engine. In the intake stroke, while the engine keeps the exhaust valve open, a part of the exhaust gas flows back to the chamber, which is called valve overlap. An advantage of internal EGR is its compactness because it works with a system of engine valves and ports instead of external devices. However, as a disadvantage, the charging efficiency at the engine intake is not sufficient and not easy to control.

The fundamental structure of an external EGR is shown in Fig. 3.2.7. The figure shows the case of a four-cycle engine, but the case of a

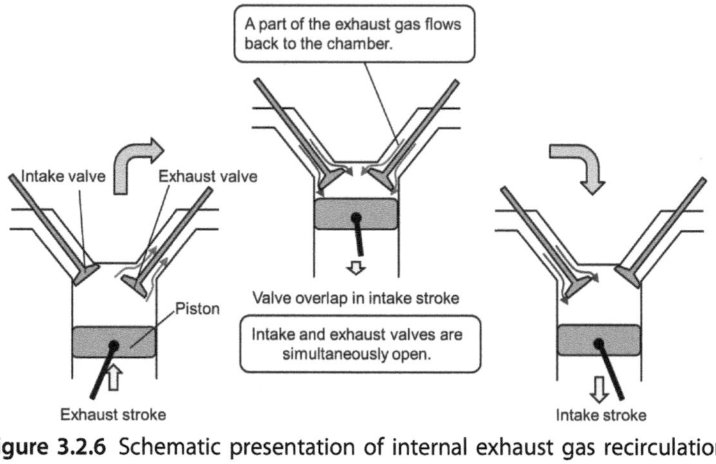

Figure 3.2.6 Schematic presentation of internal exhaust gas recirculation.

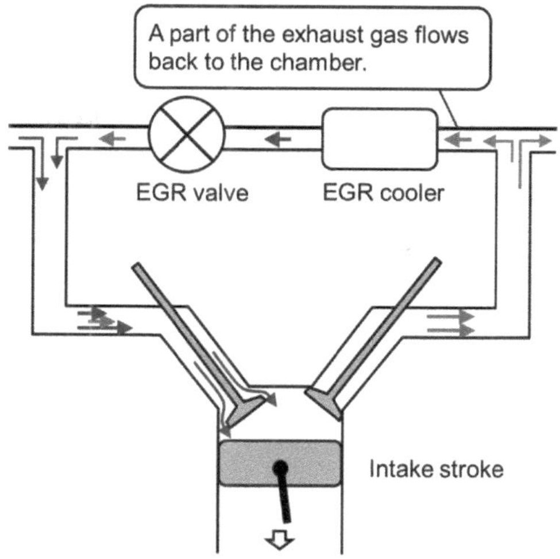

Figure 3.2.7 Schematic presentation of external exhaust gas recirculation.

two-cycle is similar. A part of the exhaust gas in the exhaust pipe with exhaust manifold is bypassed under the control of the EGR valve and flows back to the engine intake with the intake manifold. In most cases, the bypassed exhaust gas is cooled by the EGR cooler before it passes through the EGR valve to improve the charging efficiency. This feature and its controllability are the advantages of the external EGR. However, the external EGR system cannot be compact because it utilizes external devices. Moreover, an EGR cooler is equipped with a heat exchanger and a radiator.

EGR can be applied naturally to diesel engines because backpressure of exhaust gas is generally higher than the intake air pressure. However, in the implementation of high-performance EGR, several technologies have been proposed and applied. For example, an explanation of the supercharging process is carried out in the following. Providing pressurized fresh air inside the cylinder at the beginning of compression, engine power increases for the same engine dimensions and piston speed owing to charged fresh air. This is called supercharging, and methods are classified as (1) mechanical supercharging in which pressurized forced air is provided with the help of external charging pump and (2) turbine supercharging in which providing pressurized fresh air using engines exhausts turbine, which is often called turbocharging. In the external EGR applied to an engine system with a supercharger (turbocharger), depending on the position of the bypass, there exist low-pressure (LP) EGR and high-pressure (HP) EGR as shown in Fig. 3.2.8.

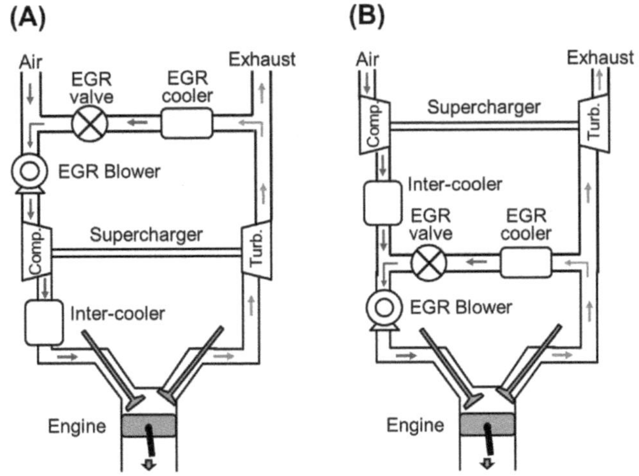

Figure 3.2.8 Schematic presentation of (A) low-pressure exhaust gas recirculation (LP EGR) and (B) high-pressure exhaust gas recirculation (HP EGR).

The figure shows the case of a four-cycle engine, but the case of a two-cycle is similar. In LP EGR as shown in Fig. 3.2.8A, a part of the exhaust gas at low pressure and low temperature is recirculated. The exhaust gas is bypassed from the downstream of the turbine to the upstream of the compressor of the supercharger. The system achieves a positive gradient differential pressure between the turbine outlet and compressor inlet. The EGR valve is installed between them. Because the recirculated exhaust gas is introduced to the inlet of the compressor, an EGR blower with lower capacity can be applied compared with that of HP EGR as shown in Fig. 3.2.8B. In general, the system is simpler than that for HP EGR. However, because the volume of the recirculated exhaust gas is large due to low pressure, the large compressor and turbine are required to improve the charging efficiency.

In HP EGR, as shown in Fig. 3.2.8B, in contrast to LP EGR, a part of the exhaust gas at high pressure and high temperature is recirculated. The exhaust gas is bypassed from the upstream of the turbine of the supercharger to the downstream of the intercooler installed downstream of the compressor. The size of devices can be compact because the volume of the recirculated exhaust gas is small due to high pressure. However, an EGR blower with high capacity is required to feed the recirculated exhaust gas to the engine intake. In general, the system becomes more precise than that of LP EGR because the recirculated gas has high pressure and high temperature.

In EGR, the ratio of recirculated gas to fresh intake air is represented by the EGR ratio. The concentration of CO_2 in fresh air intake is negligible, although the exhaust gas contains greater CO_2 as a combustion product. To measure the EGR ratio, it is sufficient to compare the CO_2 concentration of the recirculated exhaust gas with the intake fresh air. The EGR ratio, η_{EGR}, is estimated by

$$\eta_{EGR} = \frac{(C_{CO_2})_{ex}}{(C_{CO_2})_{in}} \tag{3.74}$$

where $(C_{CO_2})_{ex}$ is the CO_2 concentration of recirculated exhaust gas and $(C_{CO_2})_{in}$ is the CO_2 concentration of the whole intake gas mixture [13].

3.2.7 Other NO_x control methods

Other NO_x control technologies are shown here.

Emulsion fuel combustion

An emulsion is a dispersion of one liquid in another (e.g., water in oil) by the addition of a small amount of surfactant. There are two types of emulsion fuels. One is w/o type, where water droplets are included in the oil. The other is o/w type, where oil droplets are included in the water. Usually, w/o-type emulsions have been often used for diesel engines because their viscosity is low and they are easy to handle. When the emulsion fuel is atomized inside the higher temperature field, water in oil is instantaneously vaporized and the secondary atomization occurs to induce rapid combustion. The combustion heat is partially lost by the vaporization of the water. As a result, the combustion temperature decreases, and the suppression of thermal NO occurs. In the combustion of emulsion fuels, higher efficiency combustion is observed with simultaneous suppression of CO, soot, and NO formations. The problem with this technology is the decrease in energy efficiency due to the vaporization of water. However, the compensation is possible by the precise control of the combustion. The efficiency of NO reduction is approximately 30–40% in the emulsion fuel combustion. The cost for the preparation of the emulsion fuel is approximately 5% of fuel cost.

Water or steam injection

In this method, water or steam is injected into the combustion engine to decrease the combustion temperature and thermal NO emissions. One of the problems associated with this method is the decrease in the energy

Table 3.2.1 Various types of NO_x control technologies.

Method	Principle	Fuel	NO_x removal efficiency	Operating cost
Selective catalytic reduction	NH_3 reduction for NO_x	Gas or oil	~90%	High
Diesel particulate NO_x reduction or lean NO_x trap	Catalytic NO_x reduction	Diesel oil (automobile)	~80%	High
Exhaust gas recirculation	Temperature and O_2 decrease	Gas or oil	30–50%	High
Emulsion fuel combustion	Temperature and O_2 decrease	Oil	30–40%	High
Water or steam injection	Temperature decrease	Gas or oil	30–50%	Medium
Lower air ratio combustion	O_2 decrease	Gas or oil	10–20%	Low

conversion efficiency because part of combustion energy is consumed to the vaporization of water.

Lower air ratio combustion

Lower air combustion is to decrease NO_x with fuel-rich state combustion. This method is a simple and convenient way, but a large amount of CO and soot are generated, and fuel consumption increases.

Various types of NO_x control technologies explained here are summarized in Table 3.2.1.

3.3 SO_x control

3.3.1 Procedure for De-SO_x in the emission

The procedure for De-SO_x in the emission is presented [2,14]. The formation of sulfur oxide (SO_x) is induced mainly by the combustion of sulfur components in the fuel of marine diesel engines. Therefore, the simplest De-SO_x technology is to convert the fuel to the nonsulfur content fuel. City natural gas, LNG (liquefied natural gas), LPG (liquefied petroleum gas), and kerosene do not include sulfur components. However, for marine diesel engines in real situations, marine fuel oil or heavy fuel oil including much sulfur must be used as fuel. It is not practical to use low-sulfur oils because low-sulfur oil can be prepared for the limited number of ships. Effective De-SO_x methods include the following:

1. Desulfurization of heavy oil: To remove sulfur in heavy oil, hydrogen desulfurization with catalysts has been utilized. In this method, heavy oil is directly desulfurized or it is compressed by low pressure, and only the light part is desulfurized. The content of sulfur in diesel oil decreases every year using these pre-De-SO_x methods.
2. De-SO_x inside combustor: As a simple method for removing SO_x from combustion exhaust gas, there is De-SO_x method inside combustor in which an alkaline earth metal type desulfurizing agents such as $CaCO_3$, $Ca(OH)_2$, etc., are blown into combustion apparatus or flue duct. This method is easy to modify the facility and does not use water; therefore, it has the advantage as a dry method. However, De-SO_x performance is limited.
3. De-SO_x aftertreatment for emission: Because it is not possible to meet the industry demand of marine diesel fuels for low-sulfur oils, De-SO_x methods for emissions or exhaust gas are needed, in which SO_x induced by combustion in the engine is removed at the exhaust pipe. This is also called aftertreatment and is very important for the reduction of SO_x from marine diesel engines. The methods of exhaust gas aftertreatment have been developed for heavy-oil-fired and coal-fired power generation plants and are classified into two groups: the wet and dry methods. The wet method becomes semidry one when the temperature at the exit of the wet-type apparatus becomes higher than the dew point temperature. They are explained in the next subsection.

3.3.2 Wet methods of De-SO_x

When the fuel oil is fired, combustible sulfur in the fuel is mostly converted to SO_2, and approximately 2% of SO_2 is oxidized to SO_3. SO_3 is highly water-soluble and converted to sulfuric acid (H_2SO_4). SO_3 is almost transparent and has a pungent odor. It reacts with water and reduces the pH of water to an acidic state (pH < 7)

$$SO_2 + H_2O \leftrightarrow H_2SO_3 \tag{3.75}$$

$$H_2SO_3 \leftrightarrow H^+ + HSO_3^- \tag{3.76}$$

$$H^+ + HSO_3^- \leftrightarrow 2H^+ + SO_3^{2-} \tag{3.77}$$

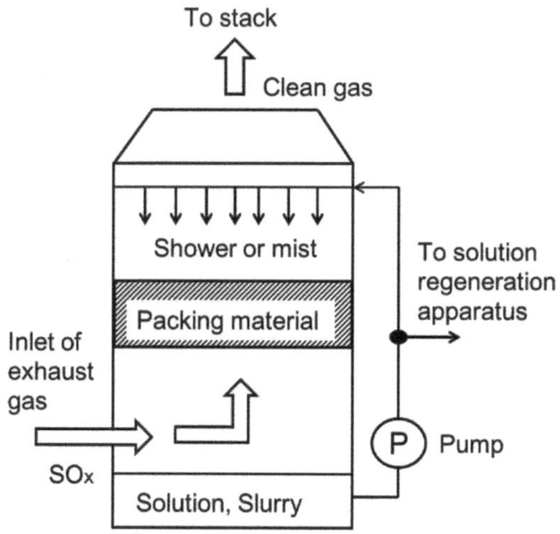

Figure 3.3.1 Typical structure of chemical absorption tower.

Alkali metal ions and alkali ions such as Na^+, K^+, Ca^{2+}, Mg^{2+}, NH_4^+, etc. can be used to fix these HSO_3^- and SO_3^{2-} ions in the solution. The soluble amount greatly depends on the pH of the solution, and pH should be kept high for effective reduction. In real aftertreatment systems, these gas—solution reactions are carried out using a wet absorption tower. A typical structure of a chemical absorption tower is shown in Fig. 3.3.1. The exhaust gas is introduced into the lower inlet. Inside the tower, there are many showers or mist nozzles, and the exhaust gas is washed by the solution. Typical hole diameter of holes on the shower nozzle is 1 mm. Gas and liquid pass through the packing material with gas—liquid contacting. Gas—liquid contact can be enhanced by the packing materials, which are not used for small towers. The chemical solution is circulated by a pump. The solution line is connected to the chemical solution regeneration apparatus, and some byproducts are processed there.

Typical flow chart of the wet scrubbing method for marine diesel engines is shown in Fig. 3.3.2. Main De-SO_x technologies including the applications to fossil fuel power generation plants are introduced. To control marine

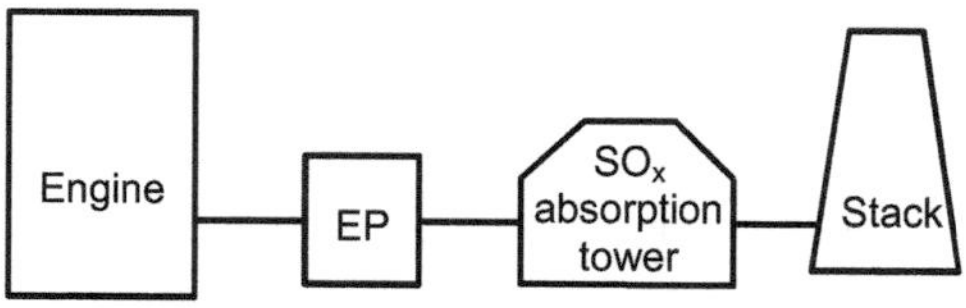

Figure 3.3.2 Typical flow chart of wet scrubbing method for marine diesel engines. *EP*, electrostatic precipitator.

diesel engine emissions by De-SO_x methods, caustic soda or sodium hydroxide (NaOH) or sodium carbonate (Na_2CO_3) solution methods are often used and explained next.

Sodium hydrate/sodium carbonate solution absorption

When water-soluble NaOH or Na_2CO_3 is used for De-SO_x, high absorption performance can be achieved with no problems of scaling. The process is shown as follows:

$$SO_2 \rightarrow \underset{\text{or } Na_2CO_3}{NaOH} \rightarrow NaHSO_3 \xrightarrow{NaOH\downarrow} Na_2SO_3 \underset{\text{Oxidation}}{\rightarrow} Na_2SO_4$$

To recover sodium sulfite (Na_2SO_3) or sodium sulfate called salt cake (Na_2SO_4) from the solution, a large amount of NaOH is required. The adsorption chemical reactions are as follows:

$$2NaOH + SO_2 \rightarrow Na_2SO_3 + H_2O \quad (3.78)$$

$$Na_2CO_3 + SO_2 \rightarrow Na_2SO_3 + CO_2 \quad (3.79)$$

In order to obtain Na_2SO_4, Na_2SO_3 should be oxidized as

$$Na_2SO_3 + O_2/2 \rightarrow Na_2SO_4 \quad (3.80)$$

In the treatment of marine diesel emissions, diluted wastewater including Na_2SO_4 is transparent and can be drained to the sea with dilution according to regional regulations for seawater. The method of conversion to the byproducts such as H_2SO_4 and $CaSO_4$ may exist but rarely used for marine diesel engines because byproducts should be avoided inside the ships from the perspective of danger and limited space.

The temperature distribution inside the absorption tower is set properly. When temperature T at the exit is more than 200°C, it can be a dry method. The solution including Na_2SO_3 and Na_2SO_4 is vaporized, and they can be recovered by the PM of Na_2SO_3 and Na_2SO_4 using electrostatic precipitator (EP or ESP) or bag filter system followed by the absorption tower. This system is called "semidry" system and often used for De-SO_x for the exhaust gas from glass melting furnace in the glass bottle production system factory [15]. However, if this method is applied to marine diesel engine systems, the generated PM of Na_2SO_3 and Na_2SO_4 should be treated either inside or outside the ship.

There are other types of the wet De-SO_x method. Fig. 3.3.3 shows a classification of the wet De-SO_x methods. In all these methods, SO_2 is initially absorbed to the solution to become $HSO_{3.}$ in the same manner.

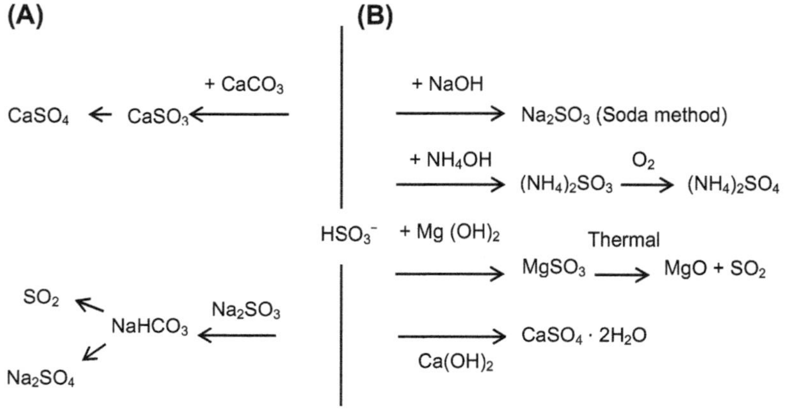

Figure 3.3.3 Various wet De-SO_x methods. (A) Reactions with carbonates and sulfates. (B) Reactions with hydroxide chemicals (alkali). *(Fig. 3.15 on page 69 of Ref. [2].)*

After that HSO_3^- is fixed with various De-SO_x chemicals. The methods are classified into the two groups: (1) De-SO_x reactions with carbonates and sulfates and (2) De-SO_x reactions with hydroxide chemicals (alkali) as shown in the figure.

Most useable method for De-SO_x of marine diesel engines is currently the NaOH method. The other methods in the figure are also explained in the next.

Ammonia solution absorption

Ammonia–water solution is used as an adsorption solution of SO_2. As a byproduct, ammonia sulfate ($(NH_4)_2SO_4$) is recovered from the solution. The reduction process is shown in the following reactions.

$$2NH_4OH + SO_2 \rightarrow (NH_4)_2SO_3 + H_2O \tag{3.81}$$

$$(NH_4)_2SO_3 + O_2/2 \rightarrow (NH_4)_2SO_4 \tag{3.82}$$

In this process, pH control is very important to obtain high De-SO_x efficiency. When pH is greater than 7, the pressure of ammonia becomes too high and large loss is induced. When pH is smaller than 5, the pressure of SO_2 gas becomes high and the removal efficiency decreases. Therefore, the process should proceed well at pH = 6.

Calcium carbonate/calcium hydrate absorption

In this method, carbonic calcium (limestone) ($CaCO_3$) or calcium hydrate (slaked lime) (Ca $(OH)_2$) powders are used as slurry to absorb SO_x with concentrations of 5−15%. The pH of the slurry is approximately 6 and SO_2 is absorbed to induce $CaSO_4{\cdot}2H_2O$. The chemical reactions are as follows:

$$Ca(OH)_2 + SO_2 \rightarrow CaSO_3{\cdot}(1/2)H_2O + (1/2)H_2O \quad (3.83)$$

$$CaCO_3 + SO_2 + H_2O \rightarrow CaSO_3{\cdot}(1/2)H_2O + CO_2 + (1/2)H_2O \quad (3.84)$$

In these reactions, $CaSO_3$ is induced. This is oxidized to induce plaster ($CaSO_4$).

$$CaCO_3{\cdot}(1/2)H_2O + (1/2)O_2 + (3/2)H_2O \rightarrow CaSO_4{\cdot}2H_2O \quad (3.85)$$

This reaction may not progress completely in the absorption tower, but part of $CaSO_3$ should be converted to $CaSO_4$ outside the tower using air oxidation apparatus such as rotary atomizers. $CaSO_4$ is separated and reused. This method is frequently used for oil- or coal-fired power generation plants on land. However, it may be limited to the application to marine diesel engines because a large amount of byproduct of $CaSO_4$ is induced, which is difficult to handle inside ships.

Magnesium hydrate absorption

Compared with NaOH, magnesium hydrate ($Mg(OH)_2$) slurry is a cheaper absorption chemical, and wastewater can be drained to the sea. Unlike NaOH, $Mg(OH_2)$ is not easily soluble in water, and it reacts with SO_2 in a slurry state at the absorption tower.

$$Mg(OH)_2 + SO_2 \rightarrow MgSO_3 + H_2O \quad (3.86)$$

The generated $MgSO_3$ should be oxidized to magnesium sulfate, $MgSO_4$, which is easily soluble in water and can be released to the sea with dilution. To our knowledge, there have been no applications of this method to SO_x scrubbers in marine diesel engines to date. However, they have the potential for future applications because of their lower cost.

3.3.3 Dry method of De-SO_x

Compared with the wet De-SO_x method, the dry one is less commercialized because of its lower SO_x removal efficiency and higher cost. The dry method also should use the adsorbents such as activated carbon, which is difficult to handle. However, it has recently been applied to a coal-fired power generation plants located near a large city to require ultraclean exhaust gas. To our knowledge, there have been no applications of the dry method in marine diesel engines to date. In the following texts, three dry methods are explained based on Ref. [2] as future procedures for marine diesel engines.

Adsorption with activated carbon

When SO_2 gas in the presence of oxygen and moisture is attached to activated carbon, SO_2 is oxidized to SO_3 on the surface of the activated carbon, which is used as a kind of gas filter. It reacts with water vapor to form H_2SO_4 in the adsorption process. Sulfuric acid (H_2SO_4) adsorbed by the activated carbon can be removed or decomposed by water washing desorption, water vapor desorption (approximately 300°C), or high-temperature reduction desorption (~370°C). Activated carbon filters are costly; therefore, more cost-effective activated carbons are needed to develop this technology. However, there are cases that it has recently been applied to a coal-fired power generation plant in Japan.

Electron beam process

In the electron beam process, high-energy electrons are applied to the exhaust gas after ammonia (NH_3) gas injection. Simultaneous removal of SO_x and NO_x is possible, and an ammonia-type fertilizer can be obtained as a by-product. Fig. 3.3.4 shows the schematic of the apparatus used to treat the exhaust gas of oil-fired boilers. The process consists of ammonia injection, electron beam application, byproducts separation apparatus, and gas-gas heater (GGH) for heat recovery. Exhaust gas from the boiler is first treated by the EP to remove fly ash. Water spraying is carried out in the cooler tower until the gas temperature becomes approximately 65°C for De-NO_x and De-SO_x. Simultaneous De-NO_x and De-SO_x can be optimally carried out at a temperature of approximately 65°C. The exhaust gas

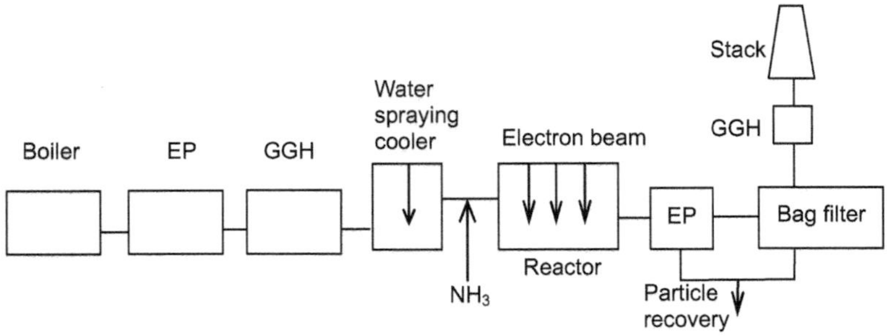

Figure 3.3.4 Process diagram of electron beam method. *EP*, electrostatic precipitator; *GGH*, gas-gas heater.

is introduced into the reactor and ammonia gas is injected. Electron beam is irradiated through an aluminum thin film using the electron beam acceleration apparatus. As a result, O_2 and H_2O are partially dissociated to produce radicals such as O, OH, HO_2, and ions such as O^- and OH^-. These radicals and ions react with coexisting SO_2 and NO_2 to form H_2SO_4 and HNO_3 mist. Furthermore, the mist reacts with the injected ammonia to generate ammonia sulfate and ammonia nitrate aerosols or particles, collected by the EP located downstream of the electron beam reactor. In 1977, a test plant that had this type of apparatus with an emission flow rate of 3000 Nm^3/h was operated, and it achieved more than 90% of SO_x removal and 70% of NO_x removal for boiler emission [16].

Pulse corona induced plasma chemical process

The pulse corona induced plasma chemical process (PCPP) method is explained based on Ref. [17]. This method was proposed by Dr. Senichi Masuda. Although the system is similar to that of the electron beam method, NTP induced by nanosecond time-scale corona discharge under high voltage (~50 kV) is used instead of the electron beam apparatus to oxidize SO_2 and NO in the emissions. In the electron beam apparatus, the energy of the accelerated electron is approximately more than 100 keV that generates X-ray, which may cause a danger to humans. However, in the PCPP method, the maximum energy of electrons is less than 100 eV which is optimum for radicals generation. Therefore, no X-ray irradiation occurs. The process of PCPP for boiler emissions is shown in Fig. 3.3.5. Firstly, the gas passes through the air heater (AH) at a temperature of 200°C. EP at the first stage can remove fly ash particles. Then, the gas is cooled down to 90°C, and NH_3 gas is injected. The gas is introduced to the PCPP reactor,

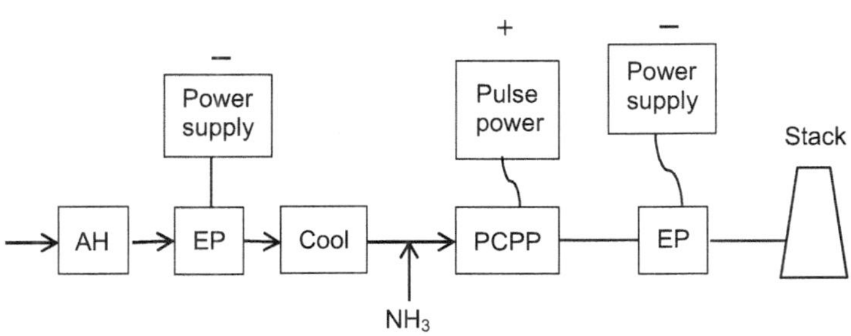

Figure 3.3.5 Process diagram of pulse corona induced plasma chemical process method. *EP*, electrostatic precipitator; *AH*, air heater.

which has a similar structure to the EP having a pair of high voltage electrodes. Secondly, high-voltage pulse corona discharge is induced in the reactor. The reduction and oxidation reactions of NO and SO_2 proceed in the presence of radicals and ions.

$$NO + N \rightarrow N_2 + O \tag{3.87}$$

$$NO + \begin{pmatrix} OH \\ HO_2 \\ O \end{pmatrix} \rightarrow NO_2 \rightarrow N_2O_5 \xrightarrow{+H_2O} HNO_3 \tag{3.88}$$

$$SO_2 + \begin{pmatrix} OH \\ O^- \end{pmatrix} \rightarrow SO_3 \xrightarrow{+H_2O} H_2SO_4 \tag{3.89}$$

It is noted that these industrial EPs are energized by the negative polarity (−) pulse power supply and the PCPP reactor is usually energized by the positive polarity (+) pulse power supply. Finally, coexisting NH_3 reacts with HNO_3 and H_2SO_4.

$$HNO_3 + NH_3 \rightarrow NH_4NO_3 \tag{3.90}$$

$$H_2SO_4 + 2NH_3 \rightarrow (NH_4)_2\ SO_4 \tag{3.91}$$

The aerosols of NH_4NO_3 and $(NH_4)_2SO_4$ are captured by the EP followed by the plasma reactor. The cleaned gas is released to the atmosphere. NH_4NO_3 and $(NH_4)_2SO_4$ particles can be reused as fertilizer. $(NH_4)_2SO_4$ decomposition occurs at the temperature more than 120°C. The following thermal reaction also occurs at a temperature less than 80°C to enhance De-SO_x.

$$2NH_3 + SO_3 + H_2O \rightarrow (NH_4)_2SO_4 \tag{3.92}$$

The PCPP method is a novel simultaneous NO_x and SO_x reduction method and has the potential of replacing the ordinary wet gas treatment techniques [2]. However, the following problems for commercial applications should be noted: (1) High operating costs (~5% of power input) for the pulse corona plasma discharge is needed. (2) High NH_3-associated cost (approximately one-third of the total cost) is needed. (3) The residence time in the reactor should be large, more than 10 s is needed.

The historical development of NTP technologies and its applications to coal-fired flue gas is reviewed by Yan et al. [18] with fundamentals and environmental applications.

3.4 Particulate matter control

There are currently no universally recognized international regulations concerning PM itself for exhaust gases of marine diesel engines; therefore, aftertreatment is not carried out. However, because black smoke released from the stack of a marine diesel engine gives a bad impression from the viewpoint of environmental protection, regional or self-regulation is implemented. A reverse cleaning regeneration system using a ceramic DPF and compressed air [19] is an example of a commercial product for removing black smoke emission. In this section, characteristics of targeted $PM_{2.5}$ is described followed by the explanations of control technology.

3.4.1 $PM_{2.5}$ and measurements

PM whose aerodynamic diameter is $d_a = 2.5\ \mu m = 2.5 \times 10^{-6}$ m or less is called $PM_{2.5}$. The definition of the aerodynamic diameter d_a is given by

$$d_a = \sqrt{\frac{18\mu u_s}{\rho_0 g}} \tag{3.93}$$

where μ is the viscosity coefficient of air, u_s is the particle settling velocity, $\rho_0 = 1\ g/cm^3$ is the unit density or density of water, and g is the gravitational acceleration. Fig. 3.4.1 schematically shows the forces acting on a single particle. The shape of the particle is considered to be ellipsoid as an example. When the particle falls down, the drag force $\boldsymbol{F}_D$ becomes equal to the gravitational force $\boldsymbol{mg}$, and the velocity of the particle reaches u_s. From Eq. (3.93), Eq. (3.94) on the particle settling velocity of the particle with d_a is obtained.

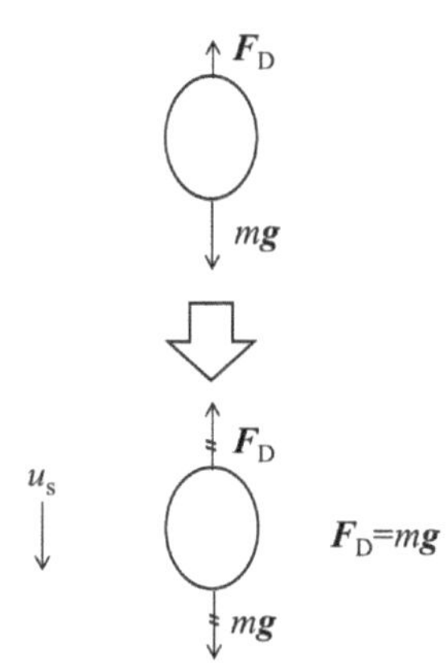

Figure 3.4.1 Description of particle settling velocity, u_s.

$$u_s = \frac{d_a^2 \rho_0 g}{18\mu} \tag{3.94}$$

On the other hand, the Stokes diameter d_{St} is defined as Eq. (3.95) and expressed with d_a,

$$d_{St} = \sqrt{\frac{18\mu u_s}{\rho_p g}} = \sqrt{\frac{d_a^2 \rho_0}{\rho_p}} \tag{3.95}$$

where ρ_p is the density of the particle. The maximum settling velocity of $PM_{2.5}$ in the atmosphere is calculated from Eq. (3.94) as

$$u_s = \frac{(2.5 \times 10^{-6})^2 \times 10^3 \times 9.81}{18 \times 1.822 \times 10^{-5}} = 187 \times 10^{-6}\ \text{m/s} = 187\ \mu\text{m/s}$$

The settling velocity of PM_{10} in the atmosphere is calculated as

$$u_s = 187 \times 4^2 = 2.990\ \mu\text{m/s} = 2.99\ \text{mm/s}$$

The settling velocity of $PM_{2.5}$ is about 0.2 mm/s, which is quite small. Thus, $PM_{2.5}$ is considered as suspended PM.

To collect $PM_{2.5}$ of various sizes, an impactor is often used. The idea of the stopping distance becomes important when designing an impactor. The stopping distance is the distance of the particle moving away from the gas path, and how much the particle moves when the direction of the gas streamlines suddenly changes when the particles are moving along the gas streamlines. For example, Fig. 3.4.2 shows a flow pattern of an apparatus called an inertial impactor for separating and collecting aerosol particles. Because the fluid ejected from the nozzle in the vertical direction changes the directions by 90 degrees, the larger particles in the fluid will not follow

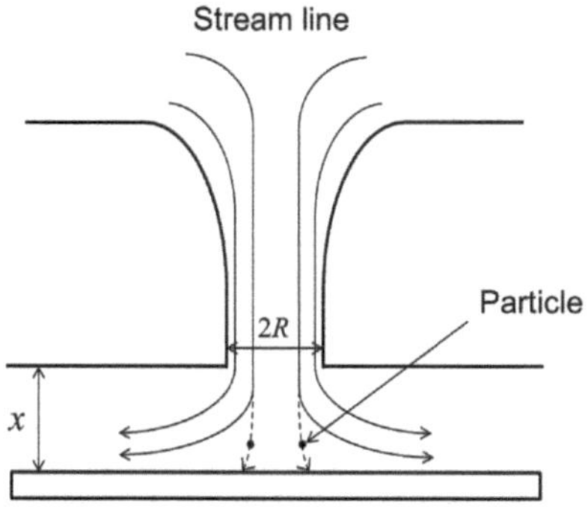

Figure 3.4.2 Schematics of an internal impactor with gap distance *x*.

the streamline because of the inertial force, while smaller particles will follow the streamline and move based on the stopping distance s in the vertical direction. Therefore, when the distance from the tip of the nozzle to the collecting plate is x, and $s > x$, the particle is captured by the plate. The critical distance s of the impactor is given as Eqs. (3.96)–(3.98).

$$\tau = \frac{\rho_p d_{St}^2}{18\mu} = \frac{\rho_p}{18\mu}\frac{d_a^2}{\rho_p}\rho_0 = \frac{d_a^2 \rho_0}{18\mu} \tag{3.96}$$

$$s = u_0\tau = u_0\frac{d_a^2 \rho_0}{18\mu} \tag{3.97}$$

$$u_0 = \frac{Q}{\pi R^2} \tag{3.98}$$

where τ is the relaxation time for the particle with the Stokes diameter of d_{St} defined by Eq. (3.95) and later given by Eq. (3.118), assuming $C_c = 1$, u_0 is the averaged flow velocity, Q is the flow rate, and R is the nozzle diameter. For example, $d_a = 2.5 \times 10^{-6}$ m, $\rho_0 = 1\ \text{g/cm}^3 = 10^3\ \text{kg/m}^3$, and $\mu = 1.822 \times 10^{-5}$ Pa·s (20°C), $Q = 1.7$ L/min $= 1.7 \times 10^{-3}/60 = 2.8 \times 10^{-5}\ \text{m}^3/\text{s}$, and $R = 0.5$ mm $= 0.5 \times 10^{-3}$ m are assumed, $u_0 = 35.7$ m/s is obtained from Eq. (3.98) and the proper gap of the impactor for $PM_{2.5}$ is calculated 680 μm as

$$s = 35.7 \times \frac{(2.5 \times 10^{-6})^2 \times 10^3}{18 \times 1.822 \times 10^{-5}} = 680 \times 10^{-6}\ \text{m} = 680\ \mu\text{m}$$

3.4.2 $PM_{2.5}$ and its components

Although $PM_{2.5}$ is often emphasized for its small size, identification of its components is important. Even if the size of the particle is small and remains in human lung cells, the risk is negligible if the component of it can be easily metabolized by humans. However, PM of heavy metal such as mercury could be dangerous to humans. Measurement examples of the elemental and ionic components of PM were reported in connection with a study on the influence of atmospheric PM on bronchial asthma attacks [20]. In this research, PM was collected at particle sizes of $< PM_{1.0}$, $PM_{1.0-2.5}$, $PM_{2.5-10}$, and $> PM_{10}$ in Himeji city, Hyogo Prefecture, Japan, for 1 week interval, and an analysis of the elemental and ionic components was conducted. The measurements were made from November 2009 to March 2010. For the measurement, a three-stage cascade impactor (critical particle

Table 3.4.1 Composition characteristics of size-resolved atmospheric particulate matter (PM) [20].

	> PM_{10}	PM_{10}–$PM_{2.5}$	$PM_{2.5}$–$PM_{1.0}$	< $PM_{1.0}$
Main elemental components	Na, Mg, Al, Si, S, Cl, K, Ca, Fe, Zn	Na, Mg, Al, Si, S, Cl, K, Ca, Fe, Zn	Na, Mg, Al, Si, S, K, Ca, Fe, Zn	Na, Mg, Al, Si, S, Cl, K, Ca, Fe, Zn
Main ionic components	Cl^-, NO_3^-, SO_4^{2-}, Na^+, K^+, Ca^{2+}	Cl^-, NO_3^-, SO_4^{2-}, Na^+, K^+, Ca^{2+}	NO_3^-, SO_4^{2-}, NH_4^+, K^+, Ca^{2+}	NO_2^-, NO_3^-, SO_4^{2-}, NH_4^+
Origin of PM generation	Most come from land	Most come from land and sea salt	Secondary generated particles in combustion-derived particles	Secondary generated particles in combustion-derived particles

sizes of 1.0, 2.5, and 10 μm) was used. Table 3.4.1 shows the measurement results. The main elements, ion components, and the formation of PM are described for each range of particle sizes measured. For $PM_{2.5}$, there are many components that are thought to contain secondary generated particles in combustion-derived particles. The elements for $PM_{2.5}$ are Na, Mg, Al, Si, S, Cl, K, Ca, Fe, and Zn, whereas NO_2^-, NO_3^-, SO_4^{2-}, NH_4^+, K^+, and Ca^{2+} ions are detected. It is possible to identify totally 26 elements of Na, Mg, Al, Si, P, S, Cl, K, Ca, Ti, Cr, Mn, Fe, Co, Ni, Cu, Zn, Se, Br, Rb, Sr, Y, Nb, Mo, Hg, and Pb. The 10 main elements are Na, Mg, Al, Si, S, Cl, K, Ca, Fe, and Zn. It is considered that the atmospheric PM does not contain obviously harmful components with high concentration, although several heavy metal components are detected. It is noted that hydrogen ion (H^+) is obviously included in $PM_{2.5}$, although it is not listed in the table.

3.4.3 Electrostatic precipitator

Principle of electrostatic precipitator

The EP is the most prevalent application of the corona as an ion source [21,22]. Although the plasma region for the direct current (DC) corona is limited in the immediate vicinity of wire electrodes, the majority of negative ions for the typical industrial EP that uses the negative polarity exist in the interelectrode space. Dust particles are charged in the corona region so effectively that the charged particles are collected on the grounded positive electrode.

An industrial EP was first developed by Cottrell in 1907 as an apparatus for collecting the mist of sulfuric acid in flue-gas emissions. Since then, precipitators have been operated at an industrial scale for collecting particulate emissions in the utility, iron/steel, paper manufacturing, cement and ore-processing industries. The EPs have greatly contributed to air pollution control efforts. The typical structure of EP is shown in Fig. 3.4.3. Corona discharge in EP is generated at high-voltage electrodes, commonly connected discharge wires, centered between flat collecting plates. The voltage for the discharge is typically 10–40 kV range and the current is 1–8 mA. The plasma occupies only a small volume near the wire; the rest of the interelectrode space is filled with ions from the corona. In this region, charging and movement of particles to the collecting plate take place. For the best dust particle–charging environment with the least power consumption, the industrial precipitator operates in the negative corona to avoid the formation of streamers and lower sparking potential associated

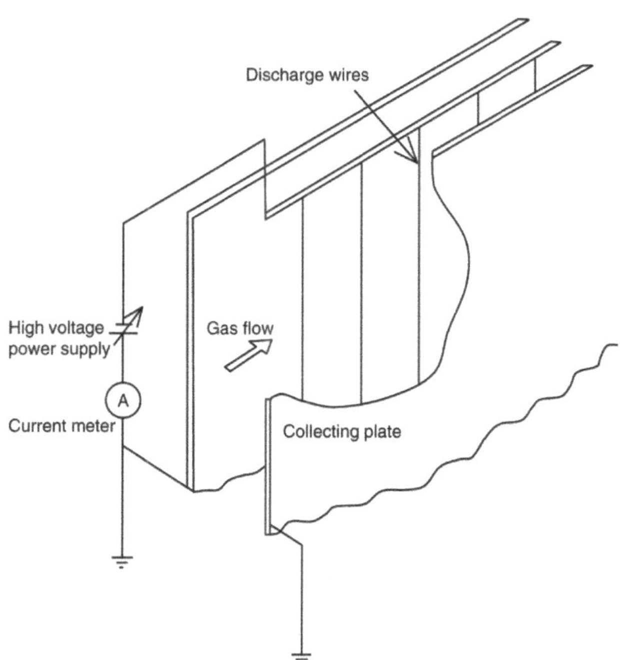

Figure 3.4.3 Typical structure of industrial electrostatic precipitator.

with them. The EP can capture nanoparticles with diameters less than 1 μm, and the pressure loss at the EP is relatively small compared to the filtration system. When the particle diameter is in the field charging regime (> 0.1 μm), the collection efficiency of the EP is proportional to the square of the electric field between electrodes. Consequently, maximizing the field strength is a major goal of the designers.

Precipitators are also used in buildings and domestic ventilation systems for the control of particles in the indoor environment. Many types of small EPs are marketed for several hundred dollars to improve sick houses. The indoor air cleaner operates with positive corona to reduce the production of ozone. The charging of particles is usually separated from the collection (two-stage EP). A typical cross-sectional structure of a two-stage EP is shown in Fig. 3.4.4. This is a commercial indoor small EP energized by a DC 5 kV power supply. In the first stage (ionizer), particles are charged, and in the second stage (collector), they are collected. The distance of the wire to the plate in the ionizer is 11.5 mm; the length of the ionizer is 18 mm; the plate-to-plate distance of the collector is 2.5 mm; and the length of the collector is 40 mm. The collection efficiency often exceeds 99.9% in this type of EP. This allows both processes to be optimized and results in somewhat better performance than that would be achieved in a standard single-stage design. The small volume of gas treated also allows the indoor cleaner to be sized more conservatively than the industrial units. The large-scale two-stage EPs are also used for fossil fuel power generation plants as a part of their emission control equipment.

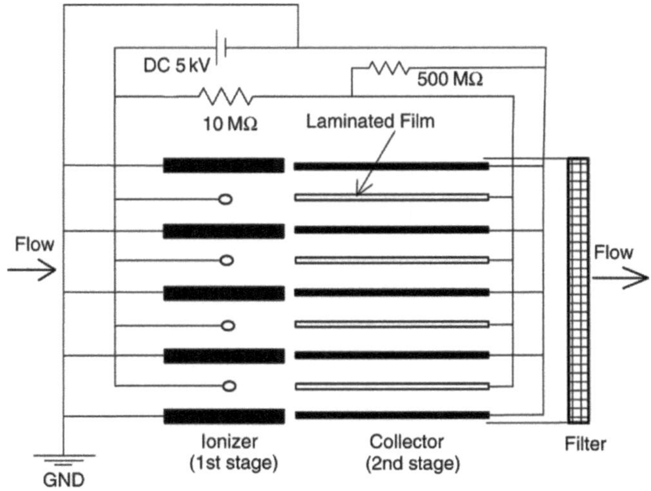

Figure 3.4.4 Two-stage electrostatic precipitator for indoor air cleaning.

Characteristics of electrostatic precipitator

EPs have several advantages and disadvantages compared to other particle control devices. Advantages include the following: (1) EPs show very high efficiency, even for very small particles. (2) They can handle very large gas volume with low pressure drop (10–20 mmH_2O) and low power consumption by fans. (3) They can handle the dry collection of valuable materials or wet collection of fumes and mists. (4) They can be designed for a wide range of gas temperatures. (5) They have low operating costs, except at very high efficiencies. (6) They have few mechanical moving parts with high reliability. The disadvantages of EPs include the following: (1) EPs have high capital costs. (2) They cannot control gaseous pollutants. (3) They are not very flexible; once installed, it is not easy to make changes in operating conditions. (4) They require a large space. (5) They show lower collection efficiency for high- and low-resistivity dust. An example of the partial collection efficiency defined as collection efficiency for each particle diameter in pulverized coal-fired combustion boilers in a stationary power plant is shown in Fig. 3.4.5. The collection efficiency is measured using three different measuring methods: cascade impactor, photo (optical) counter, and diffusion battery. The gas temperature in the EP is 160°C. The collection efficiency is at the minimum level at the particle diameter of 0.5 μm. For particles with diameters of 0.2–0.5 μm, the charging mechanism is transient from diffusion charging to field charging.

The resistance of particles is one of the main factors that have a great effect on the collection efficiency of EPs. Fig. 3.4.6 shows the relation between the electrical resistivity of the particle and the collection efficiency

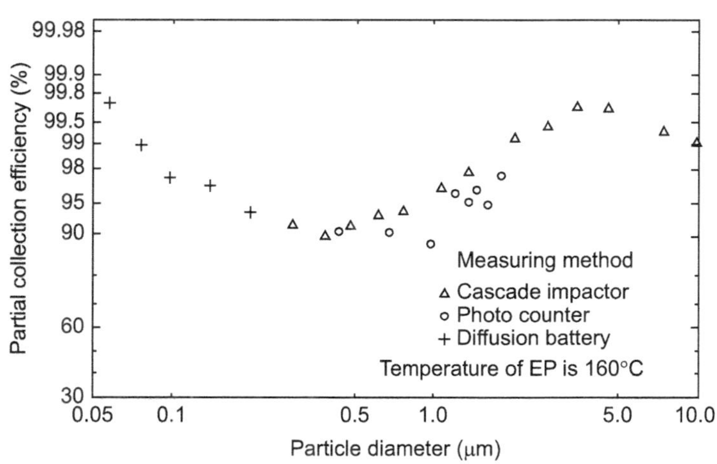

Figure 3.4.5 Partial collection efficiency for pulverized coal-fired boiler emission. *(Fig. 4.24 on page 120 of Ref. [23].)*

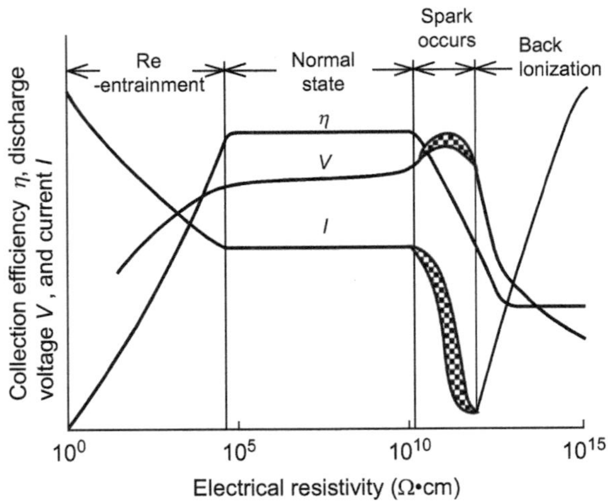

Figure 3.4.6 Relationship between particle resistance and collection efficiency.

with the discharge voltage and current. It is generally difficult for the ordinary EP to collect low-resistivity dust such as carbon particles whose electrical resistivity is less than 10^4 $\Omega \cdot$cm because the collected dust instantly loses its charge on the grounded electrode and moves into the flow field again. This phenomenon is called reentrainment and decreases the collection efficiency of the EP. It is often observed in the cleaning system of heavy-oil-fired power plants to control flue-gas emissions and in the collection of carbon soot emitted from diesel engine exhaust gas.

On the other hand, fly ash generated by the incineration of imported coal has such high electrical resistivity that streamer discharges are extended from the collected dust layer and breakdowns often occur. As a result, the collection efficiency decreases, especially, at resistivity values of more than 10^{10} Ω cm. This phenomenon is called "back corona." To avoid back corona and extend the application field of EPs, the following methods have been proposed: (1) Resistivity of the dust is decreased by temperature control (hot or cold side EP) and/or by the addition of conditional agents such as S or Na. (2) The current or charge in the dust layer is controlled using pulse high voltage power supply with a variable duty ratio. (3) The dust layer with high resistance on the collecting electrode is removed by moving brushes or rotating electrode and hoppers.

Analysis of electrostatic precipitator

Consider a dusty airflow in a rectangular channel defined by two parallel plates as shown in Fig. 3.4.7. Also, consider only the half channel between

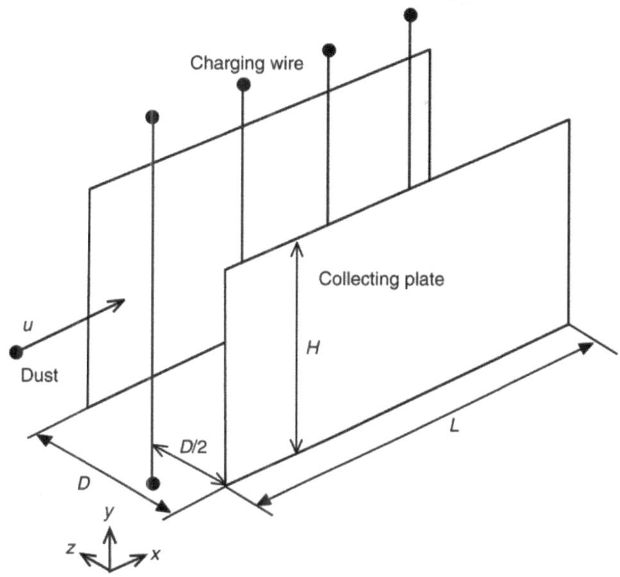

Figure 3.4.7 Analysis model for the electrostatic precipitator.

the charging wires and the plate, with width $D/2$ and height H, where D is the distance between the plates, H is the height of the collecting electrode, L is the length of the collecting plate, and u is the average velocity. With certain assumptions, the basic equation is found used in EP design, the Deutsch-Anderson equation, originally derived in 1922 [22]. The assumptions are (1) gases and particles move in the x-direction at constant velocity u, with no longitudinal mixing, (2) particle velocities are uniformly distributed in the y- and z-directions at every x location, (3) charging and collecting fields are constant and uniform; particles quickly attain terminal or settling velocity w in y-direction, and (4) reentrainment of collected particles can be negligible.

The concentration of particles will decrease with x because of the net migration of particles to the plate. A material balance on the particles flowing into and out of a very short cross section of the channel (located at x and $x + \mathrm{d}x$ and width of $D/2$) is considered. Assuming that the difference between the mass of particles flowing into and out of the slice must be equal to the mass of particles removed at the plate, the following equation is obtained.

$$uH\frac{D}{2}C_x - uH\frac{D}{2}C_{x+\Delta x} = wC_{x+\Delta x/2}H\Delta x \quad (3.99)$$

where u is the gas velocity (m/s); H is the plate height (m); D is the channel width (m); C_x is the particle concentration or loading (kg/m^3) at the position x; and w is the settling, terminal, or migration velocity in the y direction (m/s), where SI units for physical quantities are shown inside the blankets. Dividing both sides of Eq. (3.99) by Δx and taking the limit $\Delta x \rightarrow 0$, the following equation is obtained.

$$-\frac{uHD}{2}\frac{\mathrm{d}C}{\mathrm{d}x} = wHC \tag{3.100}$$

Solving this differential equation of separated variables, we obtain

$$\ln\left(\frac{C_L}{C_0}\right) = -\frac{wA_\mathrm{p}}{Q_\mathrm{c}} \tag{3.101}$$

where A_p is the area of one plane (two-sided) (m^2) (=$2HL$) and Q_c is the volumetric gas flow rate in one channel (m^3/s) (=uHD). For the whole EP, we consider A_p and Q_c to be the total collection area A and the total gas flow rate Q. Thus, Eq. (3.101) can be rewritten as

$$\frac{C_L}{C_0} = \exp\left(-\frac{wA}{Q}\right) \tag{3.102}$$

Using the usual definition of collection efficiency, the Deutsch–Anderson equation can be obtained from Eq. (3.102).

$$\eta = 1 - \frac{C_L}{C_0} = 1 - \exp\left(-\frac{wA}{Q}\right) \tag{3.103}$$

Example: Calculate the total plate area for a 98% collection efficient EP that is treating 20,000 m^3/min of air. The effective migration velocity is assumed to be 7 m/min. Further, assuming the plates are H=7 m high and L=4 m long, calculate the number of plates required.

Solution: From Eq. (3.103), we obtain A=11,177 m^2. The total area of the N collecting plates is $A = 2HL(N-1)$. Therefore, $N = (A+2HL)/(2HL) = 200.6 < 201$ plates. (Answer)

Equation of motion for a charged particle

When the EP is designed using Eq. (3.103), the migration velocity of particles should be determined [23–25] from experiments, experience, or numerical simulations. It is difficult to determine the effective migration velocity because it depends on particle size, electric field, and fuel gas characteristics. When a high-performance EP is designed, the measurement of the effective value of migration velocity should be obtained from the experiments at small-scale pilot plants. Based on this, the theoretical method to determine the trajectories and terminal velocities of a particle is explained using the two-dimensional Lagrange method in which a motion of a representative particle is considered.

Consider that a spherical particle with velocity $\boldsymbol{u}_{\mathrm{p}}$ moves into the EP at position $\boldsymbol{x}$ that consists of the discharge wire and the ground electrode. Considering the electrostatic force and the fluid resistance force as the body force, the equation that determines the motion of the particle is

$$\boldsymbol{F}_{\mathrm{e}} - \boldsymbol{F}_{\mathrm{D}} = M_{\mathrm{p}}\frac{\mathrm{d}\boldsymbol{u}_{\mathrm{p}}}{\mathrm{d}t} = M_{\mathrm{p}}\frac{\mathrm{d}^2\boldsymbol{x}}{\mathrm{d}t^2} \tag{3.104}$$

where d_{p} is the diameter of the particle; $\boldsymbol{F}_{\mathrm{e}}$ is the electrostatic force; $\boldsymbol{F}_{\mathrm{D}}$ is the fluid drag force; M_{p} is the mass of the particle; and q is the charge of the particle. The electrostatic force is expressed by

$$\boldsymbol{F}_{\mathrm{e}} = q\boldsymbol{E} \tag{3.105}$$

where $\boldsymbol{E}$ is the electric field vector that is generally the function of the co-ordinate. The fluid drag force $\boldsymbol{F}_{\mathrm{D}}$ is expressed using the drag coefficient C_{D} as

$$\boldsymbol{F}_{\mathrm{D}} = C_{\mathrm{D}}A\frac{\rho}{2}|\boldsymbol{u}_{\mathrm{r}}|\boldsymbol{u}_{\mathrm{r}} \tag{3.106}$$

where A is the projected area of a particle in the flow direction and $\boldsymbol{u}_{\mathrm{r}} = \boldsymbol{u}_{\mathrm{p}} - \boldsymbol{u}$ ($\boldsymbol{u}$ is the velocity of the fluid). The drag coefficient C_{D} depends on the particle Reynolds number of a spherical particle having an equivalent diameter d_{p},

$$Re = \frac{|\boldsymbol{u}_{\mathrm{r}}|d_{\mathrm{p}}}{\nu} \tag{3.107}$$

and is given by the following equations:

1. Laminar flow ($Re < 2$)

$$C_{\mathrm{D}} = \frac{24}{Re} \tag{3.108}$$

2. Transient flow ($2 < Re < 500$)

$$C_{\mathrm{D}} = \frac{24}{\sqrt{Re}} \tag{3.109}$$

or

$$C_D = \frac{24}{\sqrt{Re}}\left(1 + \frac{Re^{2/3}}{6}\right) \tag{3.110}$$

3. Turbulent flow ($500 < Re < 2 \times 10^5$)

$$C_D = 0.44 \tag{3.111}$$

For most aerosols with diameters, $d_p = 1-10$ μm, the following Stokes' drag force is applicable because the particle Reynolds number is much smaller than 1:

$$\boldsymbol{F}_D = 3\pi\mu d_p(\boldsymbol{u}_p - \boldsymbol{u}) \tag{3.112}$$

When the diameter of the particle is larger than the mean free path of the surrounding fluid, Eq. (3.112) is applicable to the laminar region as it is. However, the slip between the gas and the particle occurs when the particle diameter is too small (~1 μm) or in the same order as the mean free path λ. In such a case, the drag force becomes very small and Eq. (3.112) should be corrected using the correction factor $C_C > 1$:

$$\boldsymbol{F}'_D = \boldsymbol{F}_D/C_C = 3\pi\mu d_p(\boldsymbol{u}_p - \boldsymbol{u})/C_C \tag{3.113}$$

where C_C is called the Cunningham correction factor calculated by the following equation:

$$C_C = \begin{cases} 1 + 2.52\dfrac{\lambda}{d_p} & (0.01\ \mu\text{m} < d_p < 1\ \mu\text{m}) \\ 1 + \dfrac{\lambda}{d_p}\left[2.5 + 0.8\exp\left(-0.55\dfrac{d_p}{\lambda}\right)\right] & (d_p < 0.01\ \mu\text{m}) \end{cases} \tag{3.114}$$

where the value of the mean free path λ depends on the types of gas, pressure, and temperature. It becomes 6.54×10^{-2} μm for air at the one-pressure and 20°C. The values of C_C are calculated as 1.33 for $d_p = 0.5$ μm and 1.17 for $d_p = 1.0$ μm. When d_p is greater than 1.0 μm, C_C is almost equal to one. Using Eqs. (3.104), (3.105), and (3.113), the following equation is obtained:

$$\frac{d^2\boldsymbol{x}}{dt^2} = \frac{d\boldsymbol{u}_p}{dt} = \frac{q\boldsymbol{E}}{M_p} - \frac{18\mu}{C_C\rho_p d_p^2}(\boldsymbol{u}_p - \boldsymbol{u}) \tag{3.115}$$

where ρ_p is the density of the spherical particle and M_p is the mass of the spherical particle ($=\rho\pi d_p^3/6$). It is assumed that fields and particle trajectory are two-dimensional: $\boldsymbol{x} = (x, y)$, $\boldsymbol{u}_p = (u_{px}, u_{py}) = (dx/dt, dy/dt)$, $\boldsymbol{E} = (E_x, E_y)$ and air flows into the EP is uniform at a constant velocity: $\boldsymbol{u} = (u, 0)$. The x and y components of Eq. (3.115) become

$$\frac{d^2x}{dt^2} = \frac{du_{px}}{dt} = \frac{qE_x}{M_p} - \frac{u_{px} - u}{\tau_r} \tag{3.116}$$

$$\frac{d^2y}{dt^2} = \frac{du_{py}}{dt} = \frac{qE_y}{M_p} - \frac{u_{py}}{\tau_r} \tag{3.117}$$

respectively, where

$$\tau_r = C_C\frac{\rho_p d_p^2}{18\mu} \tag{3.118}$$

is the relaxation time of the particle corrected considering slips, where μ is the viscosity of the air ($= 1.822 \times 10^{-5}$ Pa·s at 20°C).

The two-dimensional trajectories (x, y) of the particles' motion can be obtained from the given distribution of q and $\boldsymbol{E}$ by integrating Eqs. (3.116) and (3.117). To obtain the distributions of q, E, and v inside the EP, the basic equations of fluid flow and electric field must be analyzed simultaneously. The details of the numerical analysis are provided in Section 4.4.

In the simplest case, Eq. (3.117) can be integrated with the assumption that the charge q and the electric field $E_y = E$ are constant. In the condition of $u_{py} = 0$ at $t = 0$, the result becomes

$$u_{py} = \tau_r\frac{qE}{M_p}\left[1 - \exp\left(-\frac{t}{\tau}\right)\right] \tag{3.119}$$

When the time reaches infinity ($t\rightarrow\infty$), the terminal velocity or the migration velocity is

$$u_{py} = \tau_r\frac{qE}{M_p} = \frac{qEC_C}{3\pi\mu d_p} \tag{3.120}$$

Eq. (3.120) gives the simplest theoretical value of the migration velocity $w = u_{py}$ used in Eq. (3.103) in the EP. Therefore, when the diameter d_p and charging quantity q of a particle are given, the electric field can be determined for the targeted collection efficiency of the EP.

Particle electric charging

The magnitude of particle charge q is greatly dependent on the method of charging. Charging by the corona discharge, which is the most popular method in EPs, consists of two types of charging mechanisms. One is fielding charging in which the monotonic ions in the corona are driven by the electric field and attached to dust particles. The other is diffusion charging, in which the induced ions are driven by thermal energy and attached to the surface of dust particles. For particles larger than 1 μm in diameter, field charging is the dominant mechanism. For particles smaller than 0.1 μm, diffusion charging is dominant. Both charging mechanisms apply to particles with diameters in the range of 0.1−1 μm.

Field charging: In each charging mechanism, the amount of particles being charged after t is expressed as

$$q_f = q_{max}\frac{t/\tau}{1+t/\tau} \tag{3.121}$$

$$q_{max} = \frac{3\pi\varepsilon_0\varepsilon_r E d_p^2}{\varepsilon_r + 2} \quad \text{(saturation charge)} \tag{3.122}$$

$$\tau = \frac{4\varepsilon_0 E}{i_c} \tag{3.123}$$

where ε_0 is the permeability in perfect vacuum ($=1/(4\pi \times 10^9)$); ε_r is the relative permeability of the particle (typical value of 5); E is the strength of the electric field (V/m); and i_c is the ion current density (A/m^2).

Diffusion charging:

$$q_d = q^* \ln\left(1 + \frac{t}{\tau^*}\right) \tag{3.124}$$

$$q^* = \frac{2\pi\varepsilon_0 d_p kT}{e} \tag{3.125}$$

$$\tau^* = \frac{8\varepsilon_0 kT\mu_i E_c}{d_p C_i e i_c} \tag{3.126}$$

where k is the Boltzmann's constant $= 1.3806 \times 10^{-23}$ J/K, T is the absolute temperature, e is the charge of an electron ($=1.602 \times 10^{-19}$ C), C_i is the root mean square of the velocity of the ion's thermal motion ($=3kT/m_i)^{1/2}$ (m/s), m_i is the mass of the ion, μ_i is the mobility of the ion (m^2/V·s), and E_c is the strength of the electric field in the charging part (V/m). Although in Eq. (3.124) the saturation charge does not exist theoretically, $q_d = 6.2q^*$ at $t = 500\ \tau^*$ can be considered as the saturation charge.

The total amount of charge given to the particle is expressed as the sum of the charging amounts in both mechanisms:

$$q = q_f + q_d \tag{3.127}$$

assuming that both mechanisms occur independently.

The charging of the particles creates a nearly immobile space charge compared with the high motility of ionic carriers. The space charge always influences corona generation, often causing the voltage to rise several kilovolts to a value more than that with no particles present at the same current. For extremely fine particulate fumes, the particulate density may be sufficient to absorb all ions before they reach the counter electrode. The EP appears to operate with no current in such a case, even though many particles are being charged and removed.

EPs have low-pressure loss and can easily collect small or nanoparticles with higher efficiency. EPs are widely used at coal-fired power plants, cement combustion plants, and iron manufacturing plants. However, the use of EP for diesel engines exhaust is limited compared with other particle collection devices. The collection efficiency decreases for low- or high-resistance particles. Further, because the exhaust gas passes through the electrically discharged region, the EP is not suitable for combustible gases.

3.4.4 Diesel particulate filter

Structure of diesel particulate filter

PM from diesel emissions primarily consists of dry soot (carbon, C), soluble organic fraction (SOF), hydrocarbon compounds, and sulfate. Typical particle composition of the emissions from a heavy-duty diesel engine tested in a heavy-duty transient cycle is 41% of carbon soot, 32% of SOF (25% of unburnt lubricating oil and 7% of unburnt fuel), 14% of sulfate and water, and 13% of ash and other pollutants (Ref. [9] in Chapter 2). Although most of sulfur (S) in the fuel is oxidized to SO_2 (gas) in the combustion process, a small fraction is oxidized to SO_3 (solid) that plays a role in the formation of sulfate in the exhaust gas. Inorganic ash is produced by the metals in the fuel and in the lubricating oil. In particular, exhaust gas from marine diesel engines operated with marine diesel oil includes more sulfate and SOF than that from marine diesel engines operated with light oil. The classification of atmospheric particles by particle size or aerodynamic diameter d_p is as follows: PM_{10} ($d_p < 10$ μm), fine particles ($d_p < 2.5$ μm), ultrafine particles ($d_p < 0.10$ μm $= 100$ nm), and nanoparticles ($d_p < 0.05$ μm $= 50$ nm). In typical diesel exhaust gas, most of the particles in mass are generated in the accumulation mode with sizes of 0.1–0.3 μm, while most of the particles in number are generated in the nuclei mode with sizes of 0.005–0.05 μm (= 5–50 nm). In the accumulation mode, the particles mainly consist of carbonaceous agglomerates and adsorbed materials. In the nuclei mode, the particles mainly consist of sulfur and volatile organic compounds (VOC). They may also contain solid carbon and metal compounds (Ref. [9] in Chapter 2).

One of the most effective tools for the accumulation of diesel PM is the DPF, which is recently used for diesel automobiles. Fig. 3.4.8 shows the structure of a ceramic wall-flow-type DPF. This type of DPF has a honeycomb structure. The flow passes through the wall of the porous filter because the plugs alternately close the passes. The effect of Brownian motion is not negligible for particles less than 0.1 μm in size. The effect becomes larger at higher temperatures for smaller particles as observed in Einstein's theory on the diffusion coefficient D:

$$D = \frac{kT}{3\pi\mu d_{\mathrm{p}}} \tag{3.128}$$

where k is Boltzmann's constant, T is the temperature, μ is the viscosity coefficient of the fluid (gas), and d_{p} is the particle diameter. DPF utilizes the effect of Brownian motion to achieve PM accumulation. Exhaust gas flows into and passes through the cells of DPF as shown in the figure. The wall of the cell plays the role of a porous filter that utilizes the Brownian motion. PM in the exhaust gas is accumulated on the wall of the cell.

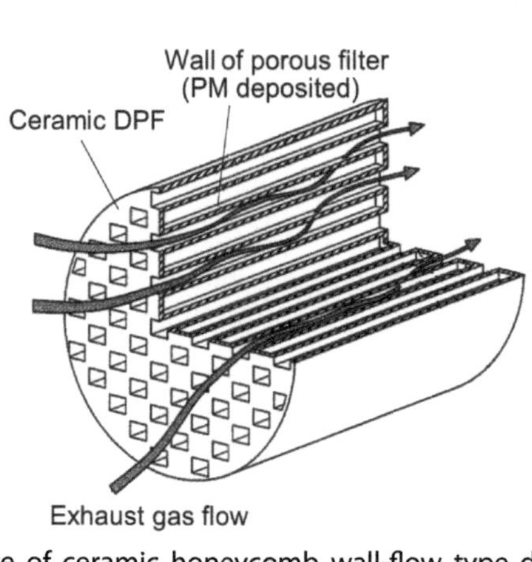

Figure 3.4.8 Structure of ceramic honeycomb wall-flow type diesel particulate filter (DPF) and exhaust gas flow paths. *PM*, particulate matter.

Performance of particle collection in diesel particulate filter

PM accumulation on a DPF is performed by three types of filtrations, i.e., sieve, cake, and deep-bed filtrations as shown in Fig. 3.4.9. Particles larger than the pore size of the filter are collected by the sieving filtration. Aggregates of particles smaller than the pore size are collected by cake filtration. Particles smaller than the pore size are collected by deep-bed filtration. Collected particles enhance the pressure drop of the DPF, negatively affecting the engine performance and creating the need to develop technologies for DPF cleaning or DPF regeneration.

In general, the collection efficiency of DPF is more than 95%. A typical DPF particle collection performance data for particles with diameters 20−320 nm are described before and after the DPF regeneration using an electric heater as follows: Before the regeneration, the number density of 6,917,749 particles/cm^3 flows into the DPF in total and 996 particles/cm^3 flow out of the DPF in total. The resulting overall PM collection efficiency is 99.986%. In the size range of 20−40 nm, the number density of particles in the exhaust gas decreases from 965,032 to 343 particles/cm^3 with a partial collection efficiency of 99.964%. In the range of 40−80 nm, the number density decreases from 2,408,606 to 166 particles/cm^3 with a partial collection efficiency of 99.993%, in the range of 80−160 nm, from 2,580,243 to 377 particles/cm^3 with 99.985%, and in the range of 160−320 nm, from 963,869 to 110 particles/cm^3 with 99.989%. After the

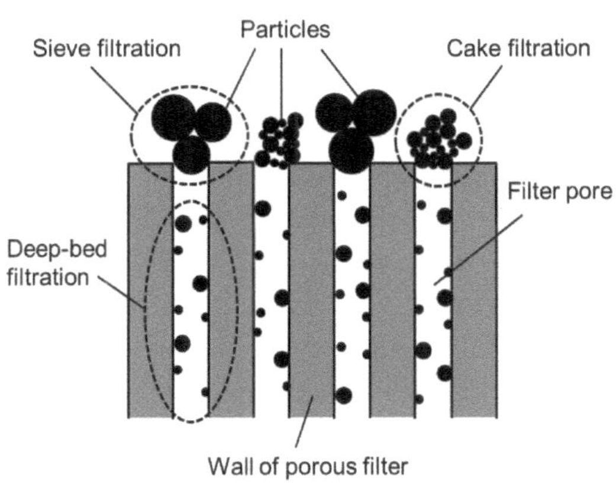

Figure 3.4.9 Representation of types of filters, i.e., sieve, cake, and deep-bed filtrations for particulate matter accumulation.

regeneration, 6,577,985 particles/cm^3 flow into the DPF in total and 1223 particles/cm^3 flow out of the DPF in total. The resulting overall PM collection efficiency is 99.981%. In the range of 20–40 nm, the number of particles in the exhaust gas decreases from 732,458 to 163 particles/cm^3 with a partial collection efficiency of 99.978%, in the range of 40–80 nm, from 2,179,589 to 352 particles/cm^3 with 99.984%, in the range of 80–160 nm, from 2,557,854 to 416 particles/cm^3 with 99.984%, and in the range of 160–320 nm, from 1,108,084 to 292 particles/cm^3 with 99.974%. It is known from the results that size-range-dependent partial collection efficiencies are almost the same before and after the DPF regeneration.

Usage example of diesel particulate filter

In recent years, most diesel automobiles in western countries are equipped with a DPF to fulfill the stringent regulations on PM emissions. DPFs require careful management; e.g., during low-speed operation of engines and/or after inappropriate manual regeneration, long-term idling is needed to regenerate DPF. On the other hand, there are no overall PM regulations for locomotive engines. However, for pollution control, some locomotives use a DPF. For marine diesel engines, recent regulations on emissions have been designed to control sulfur concentrations in fuel instead of actual PM emissions. However, regulations are expected to be more stringent in the future. To avoid the pollution of shipping products by PM emissions, DPFs for marine diesel engines are being developed.

Regeneration methods of diesel particulate filter

DPF regenerations are classified into continuous and intermittent regeneration methods. A reverse jet flow is one of the DPF regeneration techniques and represents intermittent regeneration. As an example, a high-temperature dust collector called Cerallec System is equipped with a honeycomb ceramic filter developed by NGK Insulators, Ltd., Japan, for marine diesel engines [19]. A schematic diagram of this system is shown in Fig. 3.4.10. The system is developed for a marine diesel engine. The heatproof temperature of the ceramic filter is 900°C. In this system, PM in the diesel exhaust gas is accumulated in the ceramic filter by monitoring the pressure difference of the filter. Its PM removal efficiency is 97%. When the pressure difference reaches a threshold value, compressed

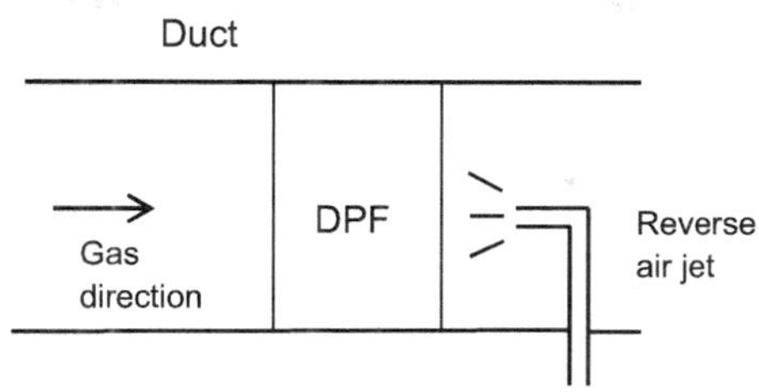

Figure 3.4.10 Reverse pulse air jet method. *DPF*, diesel particulate filter.

air is reversely injected into the filter. PM is detached from the filter by injection and collected in a dust box. This process is regularly and automatically performed. The pulse jet method is a way of injecting high-speed and intermediate-pulse air to the surface of the filter from the inside upper region of the filter.

Catalysts (Fe, Cu, or Ce) are utilized for promoting DPF regeneration at the exhaust gas temperature of 300—400°C. This method is utilized together with DPF regeneration using an electric heater. In DPF regeneration by postinjection, the temperature of DPF is increased by supplying fuel to oxidation catalysts coating DPF in oxidation reactions. Compared with O_2, free radicals are more effective in soot oxidation at low temperatures. CRT (continuous regeneration trap), which is developed by Johnson Matthey, UK, is an example system that uses free radical reactions. It uses NO_2 to generate free radicals. The technology of CRT is explained later in Subsection 3.5.3 related to Fig. 3.5.13.

The catalyst method is a continuous regeneration method and consists of two processes. In the first process, NO in the raw exhaust gas is converted into NO_2 with an oxidation catalyst installed upstream of a DPF. In the second process, soot (carbon, C) accumulated in the DPF is oxidized into CO or CO_2 with generated NO_2, and the DPF is regenerated. Although reactions can be achieved at a lower temperature of 280°C than the normal soot burning temperature of 600°C, they require the ultralow sulfur concentrations (< 50 ppm) [26] because of the catalyst degradation by sulfur and sufficient NO_x (NO_x/PM ratio should be 20 at minimum and more than 40 is preferable) condition [27]. For free radical utilization, some techniques using catalysts and NTP have also been proposed and developed.

DPF regeneration using electric heaters, which is an intermittent method, is widely used in automobiles. Soot starts to burn spontaneously at 600°C. DPF regeneration takes more than 10 min at 600°C and several minutes at > 700°C. The exhaust gas temperature is typically 500–600°C in an operation at the highest power output; thus, it is seldom possible to regenerate a DPF by heating during merely ordinal engine operation. Electrical heating is relatively easy to control DPF regeneration. DPF regeneration using electric heaters requires an electrical power of 10 kW at minimum for an exhaust gas flow rate of 1500 L/min, which corresponds to an idling exhaust gas flow from a 5-L diesel engine [28]. It requires a higher heating power for the operations with large engine loads. In automobiles, to reduce the electrical power consumption, DPF regeneration techniques have been developed using catalysts or enhancing the exhaust gas temperature in combustion and/or with heat insulation. In marine diesel engines, power consumption would be much larger with the electric heating method; thus, advanced technologies are needed.

3.4.5 Bag filter and selection of particulate matter removal systems

Bag filter

Filtering is a method to collect aerosol particles by passing the gas through textile, polymer layers, or porous media combined with sintering and separating. Precipitators based on this principle include bag filters, air filters, and ceramic filters. The advantage of filtering is the ability to collect small particles less than 1 μm in diameter. Therefore, it is effective in collecting and separating harmful high-concentration aerosols such as combustion soot and submicron fly ash from coal combustion. In this subsection, bag filters are described because they are suitable for the treatment of exhaust gases that include high-concentration aerosols from diesel engines.

Fig. 3.4.11 shows a schematic of a cylindrical-type bag filter system. Normally, exhaust gas including dust particles passes through the filters from the outside of the filters to the inside. As a result, dust particles are deposited inside of the filter material. Dust particles fall down to the hopper located at the lower region of the system using a periodical vibration applied to the filters by a supporting shaft with motors. The bag filter system experiences an increasing pressure loss as dust is collected gradually. Therefore, periodical removal of deposited particles is needed with the stimulation of the filters. The removal methods include (1) the mechanical vibration method, (2) the reverse jet method, and (3) the pulse jet method. The mechanical vibration method is the most commonly used

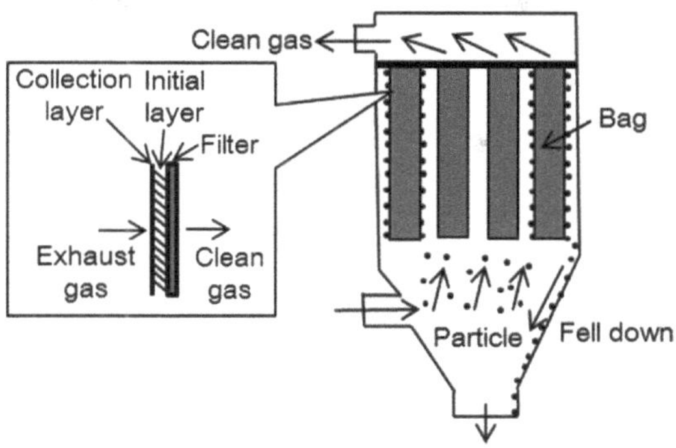

Figure 3.4.11 Bag filter [29].

technique, where the collected dust layer is broken by vibration. It is not dependent on the filter material and the exhaust gas flow rate. A reverse jet method is also frequently used. In this technology, the gas flow is temporally stopped and a reverse jet of low-pressure airflow is induced and applied to the surface of the filter in the opposite direction of the exhaust gas flow.

Table 3.4.2 shows various types of bag filter materials and properties. For high-temperature gas (> 200°C), glass fiber and PTFE known as Teflon are suitable. For erosion gases, polypropylene and Teflon are suitable. Further, the linear velocity at the surface of the bag filter should be 30 cm/s at maximum. The ordinary value is 10 cm/s because it is designed according to the pressure drop at the mechanical strength of the filter. The pressure drop at the bag filter is estimated based on the Kozeny–Carman equation

$$\Delta p = \frac{180\mu u(1-\alpha)^2 L}{g_c d_p^2 \alpha^3} \tag{3.129}$$

where Δp is the pressure drop, L is the deposition thickness, μ is the viscosity of the gas, u is the apparent mean velocity, α is the void fraction of the deposit layer, d_p is the diameter of deposited particles, and g_c is the parameter related to particle shape in the deposition layer. Δp is proportional to u and L, and inversely proportional to d_p^2. The equation is valid when the particle Reynolds number $Re_p = \rho_p d_p u/\mu < 1$. When Re_p becomes greater than 1, more loss is induced.

Table 3.4.2 Various types of bag filter materials and properties.

Type of material	Maximum temperature °C	Heat resistance	Alkali resistance	Acid resistance	Friction resistance	Liner gas velocity cm/s
Cotton	82	2	4	2	3–4	5–10
Glass fiber	257	3–4	4	4	2	5–35
Nylon	92	3	5	1	5	7.5–15
Polypropylene	92	5	5	4	4	3.5–15
Teflon	232	5	5	5	3	7.5–33
Wool	92	4	2	2	3–4	10–30

5: very strong, 4: strong, 3: normal, 2: weak, 1: very weak.
Made based on Table 4.2 on page 122 of Ref. [23].

Selection of particulate matter removal systems

Based on the characteristics of the filtering systems, the method to select the appropriate one is summarized in the flow chart shown in Fig. 3.4.12. The figure explains that firstly the averaged PM diameter d_p of the targeted gas should be measured. If it is less than 5 μm and noncombustible gas with proper resistance ($R = 10^4 - 10^{10}\ \Omega$ cm), EP should be selected. If the gas is combustible or PM does not have proper resistance ($R \leq 10^4\ \Omega$ cm or $R \geq 10^{10}\ \Omega$ cm) for EP, either bag filter or ceramic filter should be selected. Especially, the ceramic filter should be selected for high temperature ($T \geq 200°C$) and corrosive exhaust gas. Furthermore, capital and operating costs of the filters are important factors to consider in the selection process. Usually, the capital cost is in the order of EP > DPF > bag filter and operating cost is in the order of bag filter > DPF > EP.

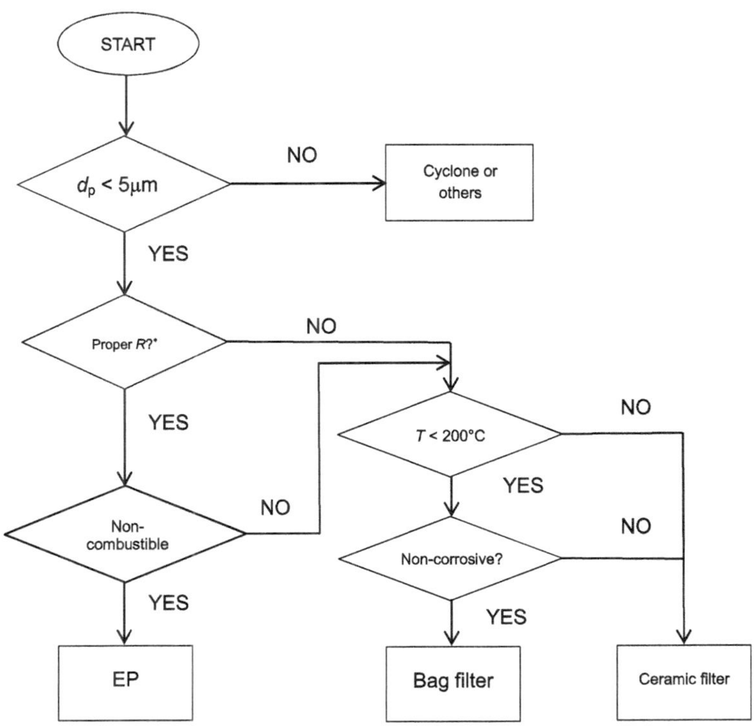

* $10^4 - 10^{10}\ \Omega$ cm

Figure 3.4.12 Selection of filtering systems. *(Fig. 4.27 on page 123 of Ref. [23].)*

3.5 Plasma treatments for NO_x and particulate matter

In this section, fundamentals of plasma or gas discharge and a new application of NTP on NO_x removal and self-regeneration of DPF are explained.

3.5.1 Fundamentals of gas discharge

To explain the plasma decomposition technology for PM deposited on DPF in diesel engines, fundamentals of gas discharge and power electronics for NTP generation should be described.

What is plasma?

Plasma is an ionization state of gas and generated by the application of high voltage between the electrodes. All substances change from solid to liquid and from liquid to gas when energy or heat is added. This change is called phase change and occurs at a constant temperature. When energy is added to the gas, electrons emerge from the neutral particles that become ions. The state in which many ions and electrons are intermingled is called "plasma" as shown in Fig. 3.5.1 [25,30–32]. The change from the gas to the plasma phase is based on the ionization reaction. The energy needed for the reaction is in the range of 1–50 eV, which is generally much more than the latent heat energy (0.01 eV). Therefore, the change from the gas to the plasma is not strictly classified as the phase change. However, plasma is often called the fourth state after solid, liquid, and gas phases of matter.

Plasma is generally defined as an ionization gas, which is electrically and macroscopically neutral (the local number density of ions n_i is equal to the number density of electrons n_e). Both ion and electron particles in plasma are moved by heat. In particular, the speed of electrons is much higher than other particles because of their small mass and mobility.

Plasma parameters

Physical quantities such as density and temperature in plasma are recognized as plasma parameters and are used to define the characteristics of the plasma. Several types of densities and temperatures exist in plasma: electron number density n_e, ion number density n_i, and gas molecule number density n_g; electron temperature T_e; ion temperature T_i; and gas temperature T_g. From the definition of the plasma, n_e is usually equal to n_i. However, the electron temperature T_e is often higher than the ion and gas temperatures T_i and T_g, whereas T_i is nearly equal to T_g. Therefore, this state corresponds to

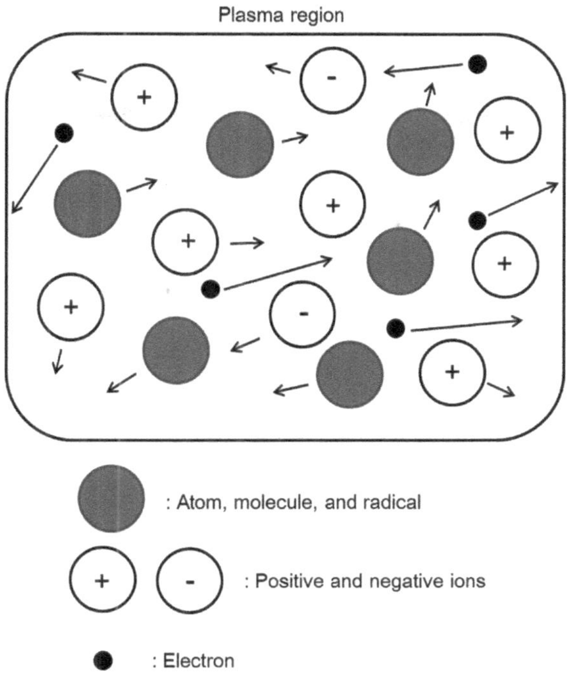

Figure 3.5.1 Various types of particles (atoms, molecules, radicals, ions, and electrons) and collisions in the plasma space.

nonequilibrium plasma state. In general, when T_g is much lower than the combustion temperature ($T_c \approx 1000°C$), the state is called "nonthermal plasma (NTP)." Because the temperature of the plasma is defined only when the energy of each particle follows the Maxwell distribution, the energy distribution function of the particles in the plasma $f(u)$ is also very important in characterizing the plasma. The definition of the plasma temperature is slightly complicated and is not a main topic in this book. Therefore, it is explained in A.3 of Appendix.

The Debye length, plasma angular frequency, the thickness of ion sheath, and plasma space potential are also important parameters for describing the state of the plasma. Another important plasma parameter is density. Because plasma is a collection of charged particles moving randomly, it can be considered a continuous medium or fluid except for the rarefied state. Therefore, the field of a quantity can be defined from an average value in a small volume, which is small macroscopically but includes sufficient number of particles. Density is the number of particles, which exist in a certain unit volume at a certain moment; therefore, ion density, electron density, and neutral particle density (gas density) can be defined. The unit of density is cm^{-3}. The electron number density of the plasma induced by low-pressure (approximately several 10 Pa) glow discharge has been investigated in detail. The electron density can be varied in the range of 10^8-10^{11} cm^{-3} by adjusting the electric discharge current from several milliamperes to several amperes. The number density of electrons is almost proportional to the electric current. On the other hand, the electron number density of the plasma in the ionosphere is around 10^4-10^6 cm^{-3}, smaller than that of the glow discharge. Further, an ionization degree is defined as the ratio of ion number density to gas-particle number density.

Classification of plasma

Plasma exists around us in nature in a variety of forms such as thunder, sun, aurora, ionosphere, stars, and in human environments such as arc welding, electrical discharge machining, flame, fluorescent light, and neon bulbs. There are several ways to classify plasma. One is based on the ionization degree. The ionization degree is defined as the number density of ions divided by the number of neutral particles. All neutral particles are ionized in the plasma when the ionization degree is 1. On the other hand, no neutral particles are ionized when the ionization degree is zero. The plasma is classified according to the difference in ionization degrees: (1) Fully ionized plasma: The ionization degree is $> 90\%$. Neutral particles have little effect on the plasma. (2) Weakly ionized plasma: The ionization degree is $<1\%$ and the effects of electrons are dominant. (3) Partially ionized plasma [31,32]: It has the middle characteristics of fully ionized plasma and weakly ionized plasma. Although partially ionized plasma and weakly ionized plasma occur in many cases, fully ionized plasma is artificially formed in a few cases; e.g., nuclear fusion. In nature, the full ionization plasma of hydrogen occupies most part of the space, for example, inside the sun.

Plasma is also classified according to the equilibrium state. The equilibrium is defined as the state in which the temperatures of electrons, ions, and neutral particles are equal and the ionization equilibrium is established. In nonequilibrium plasma, the electron temperature is usually more than the neutral particle temperature. (1) Equilibrium plasma, e.g., the atmospheric pressure arc discharge thermal plasma. (2) Nonequilibrium plasma, e.g., the low-pressure glow discharge plasma, the atmospheric pressure high-voltage pulse plasma.

Plasma is also classified depending on whether the gas temperature of the plasma T_g is as high as the combustion temperature ($T_c > 500°C$) or not. (1) Thermal plasma, $T_g \geq T_c$, e.g., plasma spraying arc discharge, flame, and plasma for garbage incineration. (2) NTP, $T_g < T_c$, e.g., discharge lamp, plasma display monitor, corona, and exhaust gas treatment plasma which is treated in this book.

Characteristics of plasma and its applications

Plasma differs from unionized ordinary gas by the following characteristics: (1) higher chemical activation, (2) higher electrical conductivity, (3) higher energy and temperature (gas or electron), and (4) strong electromagnetic radiation and light emission.

In characteristic (1), the plasma produces radicals easily when the reactive gas is inserted into it, and an extremely reactive state can be obtained with the high energy of the plasma. The fields of plasma chemistry, plasma environmental improvement, and plasma processing such as chemical vapor deposition use these characteristics. Most applications are classified into this group and used in this practical aspect. In characteristic (2), the plasma is affected by the electric and magnetic field, and the electric current can pass through the field because many electrically charged particles exist in the plasma such as ions and electrons. Using this feature, plasma parameters can be controlled and the plasma can be heated to a high temperature. There is a possibility to control the heat transfer of the plasma by the electromagnetic field. In characteristic (3), the plasma at 10–100 million °C will make a nuclear fusion reaction possible. This type of energy source is expected to be used in the future. Moreover, materials can be heated rapidly to high temperatures using the 10,000°C plasma. At lower temperatures, plasma is used in mechanical processes such as plasma welding, cutting, electrical discharge machining, and plasma spraying. In characteristic (4), many types of strong electromagnetic waves are emitted from the plasma because of high-temperature ionization. The plasma can be used as a light source, a display, and optical analysis.

3.5.2 Fundamentals of power electronics for atmospheric plasma generation

To generate plasma, electrons must be removed from the particles in neutral gas. The energy required for electric discharge is generally used to ionize the gas. There are many other methods to generate the plasma, as described in the following parts.

Steady direct current discharge plasma

The DC discharge is induced by applying DC voltage between the electrodes and ionizing the gas by breaking down the gas insulation. The DC electrical discharges are classified mainly into the corona discharge, the glow discharge, and the arc discharge ordered by the degree of current. Spark is another type of electrical discharge, which generates the plasma within a short period of time.

Corona discharge plasma

The corona discharge consists of relatively low-power electrical discharges that take place at or near atmospheric pressure. The corona is invariably generated by strong electric fields associated with small diameter wires, needles, or sharp edges from the electrodes. It is usually induced at high voltages of several or several 10 kV. The current is usually in the range of $1-100\ \mu A/cm^2$. Corona takes its name from crown which originates from mariners' observation of discharges at their ships' masts during electrical storms. The corona appears as a faint filamentary discharge radiating outward from the discharged electrode. The plasma region for the corona is limited in the immediate vicinity of electrodes.

Because the corona is relatively easy to establish, it has wide applications in a variety of processes. Applications of corona discharge processes have existed for more than 100 years, dating to the first EP. Since then, corona has been extensively used in several commercial applications and is gaining attention for use in others.

Two primary conditions for corona generation are the following: (1) The distribution of the electric field is nonuniform and its gradient (dV/dr) strength is more than the corona initiation electric field. (2) The maximum gradient of the electric potential (dV/dr) decreases by the generation of the corona.

In a nonuniform electric field, the strength of the electric field is at its maximum in the space near the electrode; therefore, the corona is induced there first. The corona near the anode is called positive corona and the corona near the cathode is called negative corona. For a typical EP employing the negative corona, positive ions are produced in the vicinity of the high-voltage electrode, whereas the negative ions exist mostly in the interelectrode space.

Corona discharge exists in several forms, depending on the polarity of the field and the electrode geometrical configuration. For positive corona in the needle-plate electrode configuration, discharges start with a burst

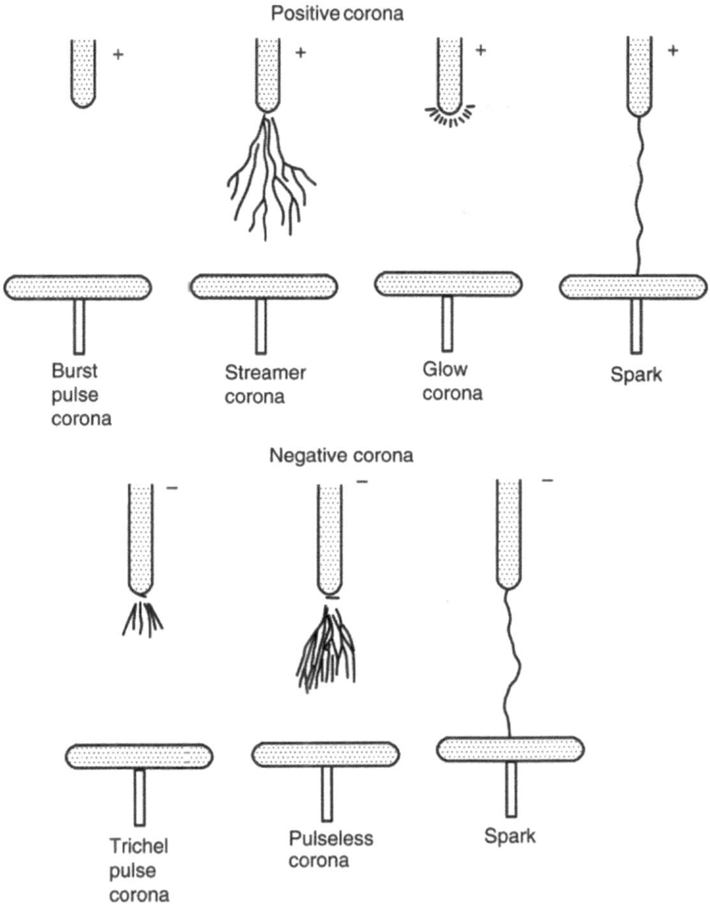

Figure 3.5.2 Various types of discharge in positive and negative coronas [33].

pulse corona and proceed to the streamer corona, glow corona, and spark discharge as the applied voltage increases (Fig. 3.5.2). For negative corona in the same geometry, the initial form will be the Trichel pulse corona, followed by pulseless corona and spark discharge as the applied voltage increases.

Alternating current or pulse corona discharge plasma

There are other methods to generate the plasma by intermittent discharge or AC discharge. Low-frequency AC discharge behaves like the DC discharge. Lately, such plasma is widely used in the applications of environmental protection. A variety of plasma reactors have been investigated by many researchers, and typical reactors and discharges are shown in

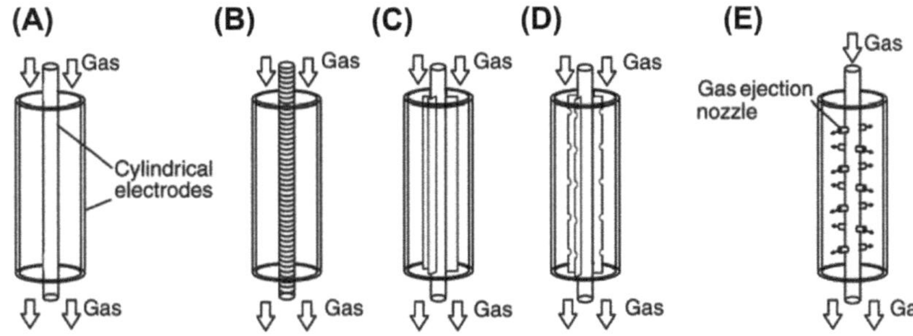

Figure 3.5.3 Various kinds of coaxial-type nonthermal plasma reactors. (A) Standard type, (B) bolt-shape, (C) vane-type, (D) scallop-type, and (E) corona-shower-type plasma reactors.

Fig. 3.5.3. Fig. 3.5.3A–E shows several types of coaxial-type plasma reactors. A high-voltage AC or pulse voltage is applied between the outer and inner electrodes in which the plasma is generated and hazardous air pollutants are decomposed. Fig. 3.5.3A is the most standard coaxial-type plasma reactor. Fig. 3.5.3B shows a bolt-shape reactor. The plasma is generated from the sharp edges of the bolt. Fig. 3.5.3C and D shows the vane-type and scallop-type center electrodes, respectively, in which the plasma is generated from the sharp edges of these electrodes. Fig. 3.5.3E shows the corona-shower-type electrode in which the treated gases are injected out from the center electrode nozzle. Fig. 3.5.4 shows the photos of the streamer corona for the wire-type electrode (wire-to-cylinder distance is 10 cm), similar to Fig. 3.5.3A. The time-dependent streamer propagation is initiated from the center electrode and moves away from the wire electrode. After 150 ns, streamers reach the ground electrode. The speed of streamer propagation is estimated as 7×10^7 m/s from these photos. When hazardous gas passes through these reactors, it is decomposed and treated to clean the gas.

Pulse power supply

It is possible to make the high-voltage pulse with short width and high frequency by turning the current on and off at the primary winding of the transformer using semiconductor switching elements. Rapid switching of high voltage is important because the strong nonequilibrium plasma can be generated with small power consumption by fast switching with a small rise time for generating high-speed electrons. If we were to raise the voltage using a pulse transformer with a large ratio of windings (secondary/primary), a sharp increase in the voltage could not be obtained. Recent developments in semiconductor switching elements make the switching voltage higher.

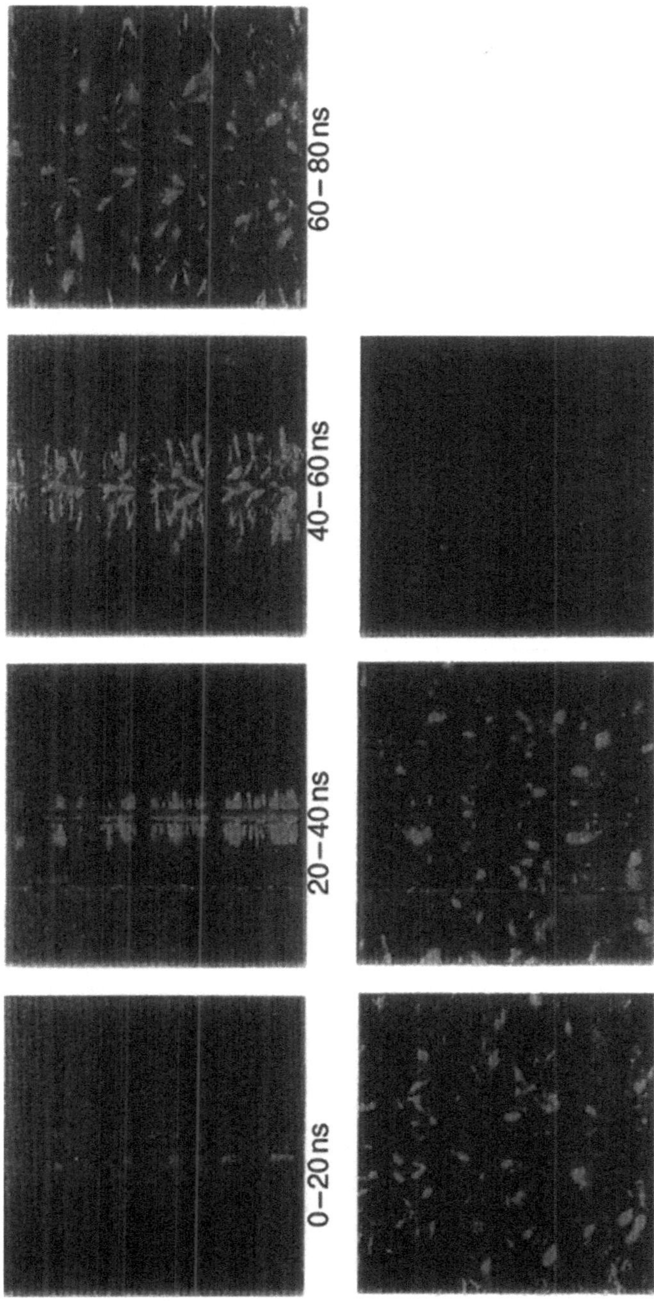

Figure 3.5.4 Photographs of the discharge structures in a coaxial-type pulse corona plasma reactor.

Figs. 3.5.5 and 3.5.6 show the examples of such types of electric circuits for the pulse high-voltage power supply. Fig. 3.5.5 shows a circuit using static induction (SI) thyristor switching. In this circuit, after the 3 kV DC voltage is switched to induce the sharp rising pulse, it is boosted by a pulse transformer (1:15) up to around 45 kV. Fig. 3.5.6 shows the circuit using the IGBT (insulated-gate bipolar transistor) elements switching. In this circuit, the DC voltage of 1.5–1.9 kV is switched to induce the sharp rising pulse and is raised by a pulse transformer (1:20) up to 35 kV. Fig. 3.5.7 shows an example of a voltage and current waveform induced by the SI thyristor electric circuit as shown in Fig. 3.5.5. The rise time of the pulse voltage is approximately 100 ns. To obtain the pulse with a smaller rise time, the cost for the power supply becomes higher.

Fig. 3.5.8 shows an electrical circuit for high-voltage pulse power supply using a mechanical switch called rotating spark gap high-voltage switching with a plasma reactor. In the figure, the rotating switch is rotated by an electric motor and a matching capacitor is chosen to adjust the electrical impedance of the reactor and the circuit. Because the mechanical switching of the secondary or high-voltage line is possible, a very sharp pulse with a small rise time (approximately several ns) can be generated as shown in Fig. 3.5.9. Furthermore, the cost of this apparatus is not very high and commercialized. However, there are several problems associated with this method: (1) The lifetime of the contact point is short. (2) The width of the pulse is relatively large. (3) It is difficult to generate a pulse with large amplitude and frequency. (4) A large space is required. (5) Electromagnetic noise emission must be considered.

3.5.3 Ozone synthesis and application to particulate matter removal

Principle of ozone generation

The ozone generation system by corona discharges was developed by Ernst Werner von Siemens approximately 100 years ago. A silent discharge is frequently used for the ozone generation system or ozonizer. As shown in Fig. 3.5.10, the silent discharge corresponds to a capacitor that has two plate electrodes, one of which is covered by dielectric materials such as glass. The silent discharge is also called dielectric barrier discharge. The ozone generation system is one of the famous applications of DBD. In ordinary ozonizers using the silent discharge, the gap distance is 2–3 mm. Ozone generation has improved greatly by lowering the gap distance to 0.05–0.2 mm [34].

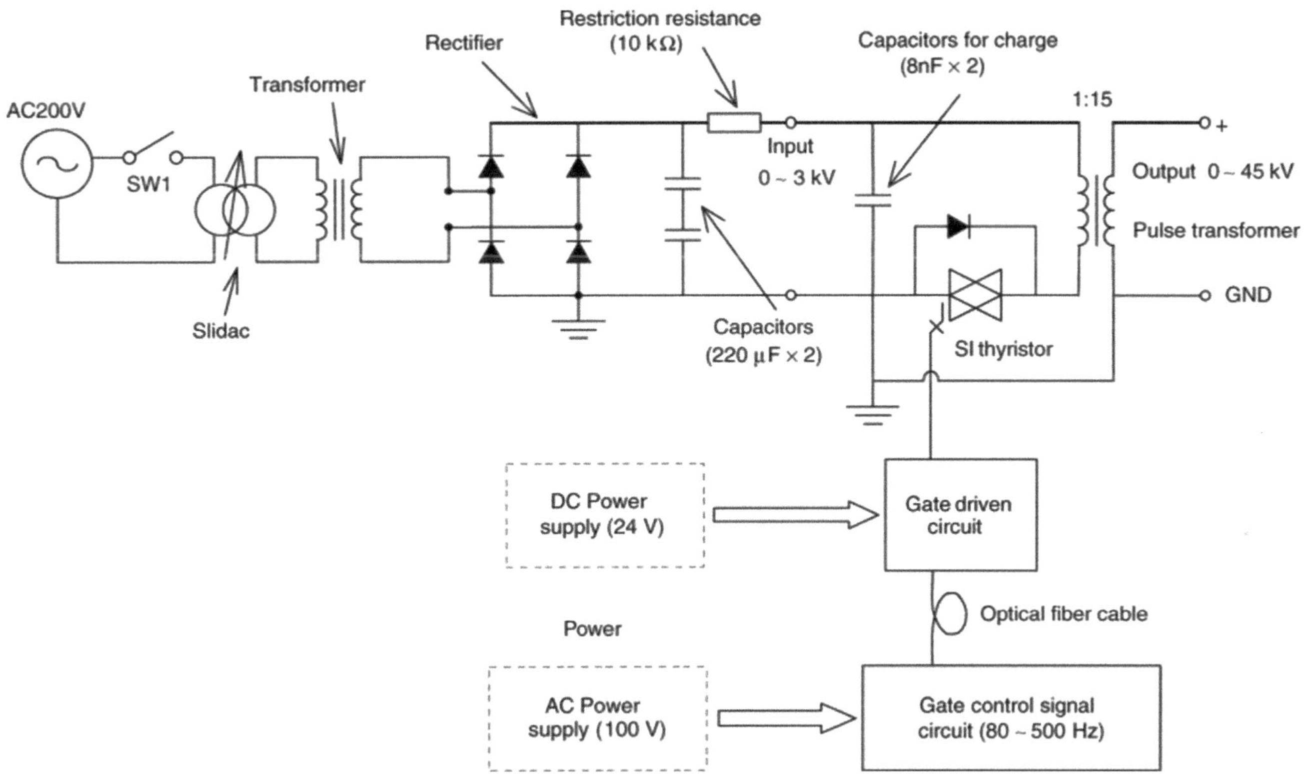

Figure 3.5.5 Electric circuit for high-voltage pulse power supply using static induction thyristor switching.

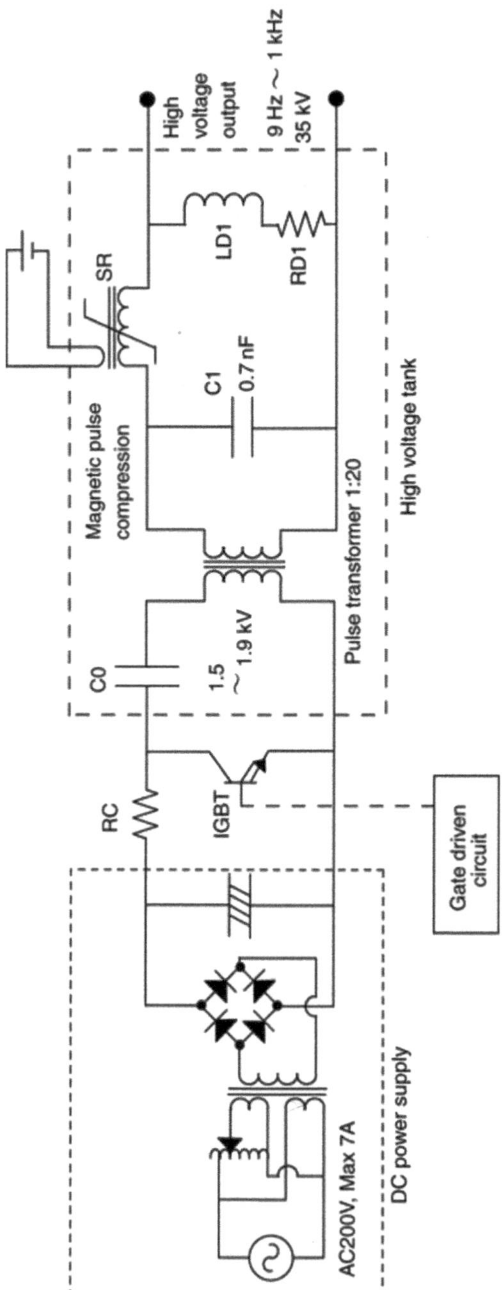

Figure 3.5.6 Electric circuit for high-voltage pulse power supply using insulated-gate bipolar transistor switching.

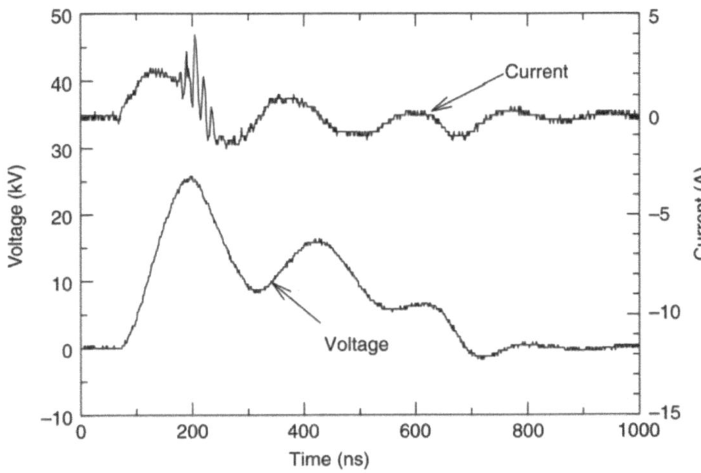

Figure 3.5.7 Voltage and current waveforms induced by the pulse power supply using static induction thyristor switching ($f = 420$ Hz).

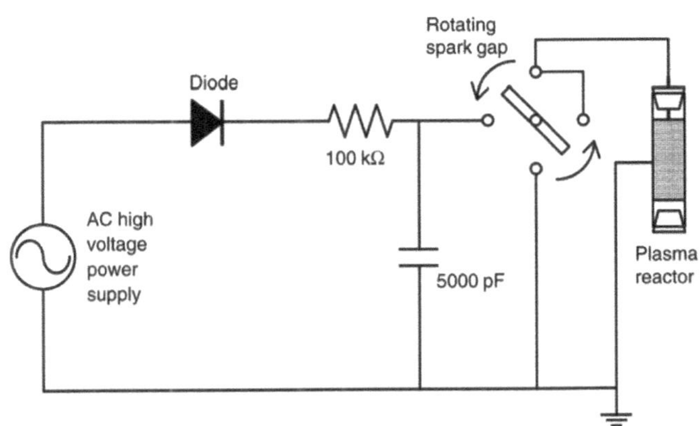

Figure 3.5.8 Electric circuit for high-voltage pulse power supply using rotating spark gap high-voltage switching.

To achieve higher efficiency, a surface discharge-type ozonizer has recently been developed. Fig. 3.5.11 shows an example of high-concentration ozone generating equipment adopting the principle of the surface discharge-induced plasma chemical process (SPCP). This SPCP ozonizer (OZS−HC−70HP, Masuda Research, Inc.) is built by attaching the evaporator of the heat pump circuit directly to the outer circumference of a cylindrical creeping discharge-type ceramic ozone generating pipe. It has a structure that can efficiently remove the heat generated in the ozone generating pipe. It is an ozone generating system that compactly

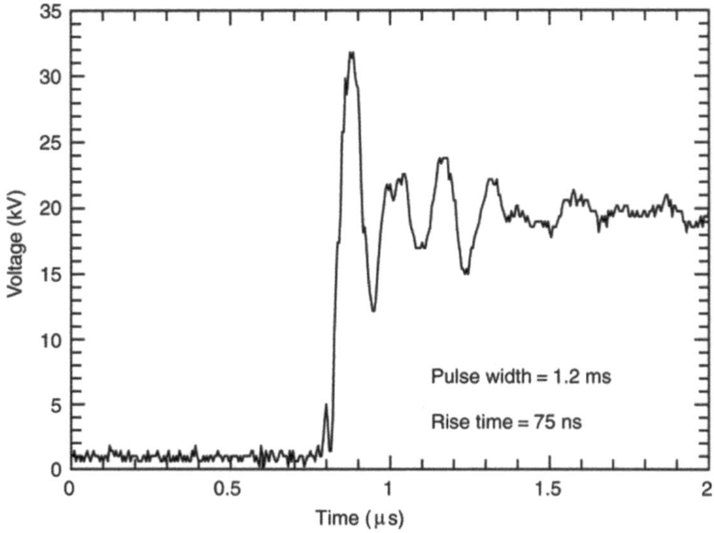

Figure 3.5.9 Voltage waveforms induced by the pulse power supply using rotating spark gap switching ($f = 210$ Hz).

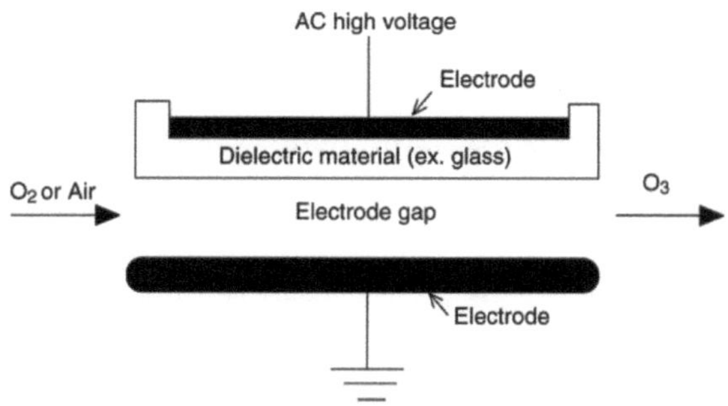

Figure 3.5.10 Structure of silent discharge ozonizer electrodes.

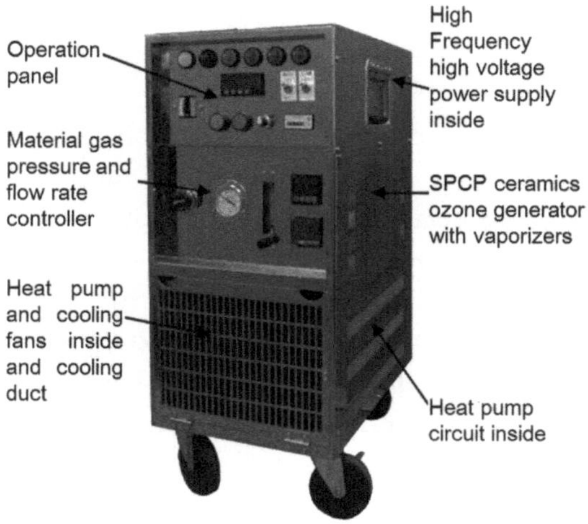

Figure 3.5.11 Commercially used ozonizer cooled by heat pump system [35].

incorporates the ozone generating pipe with the compressor, the condenser, and the evaporator of the heat pump circuit and the compact high-frequency, high-voltage power supply (HC−70, Masuda Research, Inc.) in the stainless steel housing. It is possible to control the pressure and flow rate of raw material gas (O_2) and adjust the amount of ozone generation. The system operates as follows: supply raw material gas (oxygen), connect the power supply, switch it on, and obtain ozone immediately. Compared with the water-cooled type SPCP ozonizer, the heat pump direct cooling-type SPCP ozonizer can maintain the cooling temperature at a constant low value regardless of the outside air temperature; thus, the ozone generation performance is greatly improved. Power supply is 200 VAC and raw material maximum gas flow rate is 4 L/min. There are many other applications of the system in semiconductor manufacturing, surface treatment, exhaust gas oxidization, medicine, and food processing.

Fig. 3.5.12 shows the specifications of the ozonizer and the standard ozone generation performance. The horizontal axis represents the flow rate of the supplied oxygen gas, the left vertical axis represents the ozone

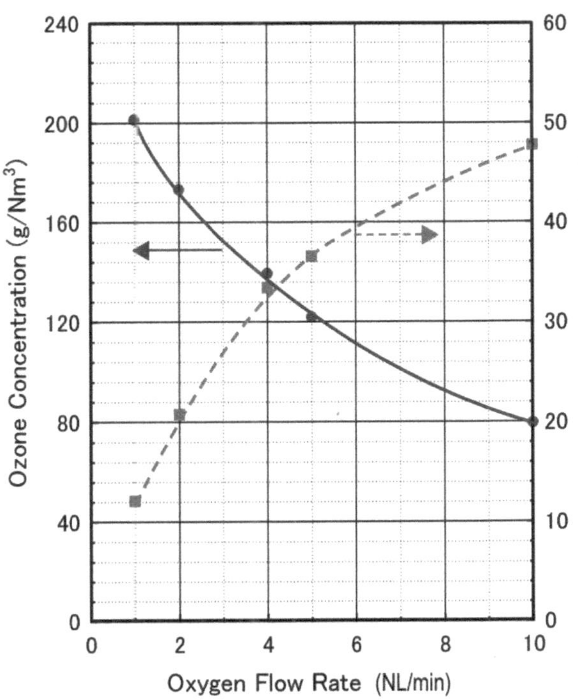

Figure 3.5.12 The relation between ozone concentration and oxygen flow rate [35], showing performance of the ozonizer.

concentration (unit: g/Nm^3), and the right vertical axis is the ozone mass flow rate (unit: g/h).

Commercial ozonizers usually have an energy yield of 90 and 150 g/kWh for dry air and pure oxygen gas, respectively. Because the theoretical limit of energy yield is 1200 g/kWh, 92.5% of the energy in a silent discharge is lost as heat. Ozone generated in a discharge is a two-step process described as follows:

Generation of oxygen radical by oxygen-electron collision processes:

$$O_2 + e \rightarrow O_2^+ + 2e, \quad \text{direct ionization} \tag{3.130}$$

$$O_2 + e \rightarrow O^+ + O + 2e, \text{ dissociative ionization} \tag{3.131}$$

$$O_2 + e \rightarrow O + O + e, \text{ dissociation} \tag{3.132}$$

$$O_2 + e \rightarrow O^- + O, \text{ dissociative attachment} \tag{3.133}$$

Generation of ozone by free radical reactions:

$$O + O_2 + M \rightarrow O_3 + M \tag{3.134}$$

where M indicates the third body such as O_2 or N_2 in the air. Reaction (3.130) is induced by a number of electrons with energy levels higher than 12.2 eV. Reactions (3.131)–(3.134) occur by the electrons with energy levels of 5–8 eV. Except for dissociative attachment, reactions in the first step (reactions (3.130)–(3.133)) are not significantly influenced by gas temperature. However, the reaction rate of the second step (reaction (3.134)) appears to be significantly influenced by gas temperature; reaction rate, hence, ozone generation rate, decreases with increasing gas temperature; however, ozone loss processes are significantly enhanced by increasing the gas temperature. Therefore, if an ozonizer is operated in lower temperature conditions, a substantial enhancement of ozone generation can be expected.

Characteristics of diesel exhaust gas and treatment methods

Diesel engine hazardous emissions are roughly classified into two groups: PM or diesel exhaust particle (DEP) and gaseous emissions. These two categories can be divided into further subgroups with the following characteristics:

PM or DEP:

1. Carbon soot particles are usually called black smoke. Carbon is the main component of soot and its incineration is very difficult.

2. SOF components are soluble in organic solvents. Because these components remain in the vapor phase at high temperatures, they can be incinerated using oxidization catalysts. However, these components increase the diameter of carbon soot particles when the temperature of the emissions decreases.
3. Oil ash is the residual component after the incineration of engine oil. It is generated mostly by deterioration and aging of the engine.
4. Mist of sulfate or sulfuric acid (H_2SO_4) is generated from gaseous emissions of SO_2 by catalytic oxidization systems with moisture causing acid rain.
5. Mist of nitrate or nitric acid (HNO_3) is generated from gaseous emissions of NO nitrate by catalytic oxidization systems with moisture causing acid rain.

Gaseous emissions:

1. Carbon dioxide (CO_2) is always produced by incineration/combustion and has a significant GWP.
2. Carbon monoxide (CO) is a hazardous emission component.
3. Nitrous oxide (N_2O) has a higher GWP than CO_2.
4. Oxides of sulfur (SO_x) are generated from sulfur in light fuel in diesel engine combustion. In general, the concentration of SO_2 among other SO_x is the highest but can be reduced by predesulfurization of fuel.
5. Oxides of nitrogen (NO_x), in general, NO, are the most dominant NO_x compound. NO_x causes acid rain, photochemical smog, and environmental and health problems.
6. Hydrocarbon compound is a general term for organic compounds (chemical compounds of hydrogen and carbon). They cause acid rain and photochemical smog. Most of them are volatile at high temperatures.

Among these components, black smoke (soot) has attracted particular attention because it is visible in the emissions from the exhaust pipes of diesel vehicles. Soot is harmful to human health. In the following, the treatment methods are described in relation to automobile engines, but they are also useful for marine diesel engines. The regulations regarding the installation of DPFs begin to be imposed in big cities throughout the world. The most typical DPF structure is a wall filter-type DPF made from honeycomb ceramics (cordierite or SiC). The flow passes through the wall of the porous filter because the plugs alternately close the passes. Carbon particles are captured on the wall of the porous filter. Other types of DPF include fibrous filter-, mat-, foam-, and multimetal plate-type DPF.

The captured soot particles must be removed periodically because the increase in the pressure of the exhaust gas decreases the engine power and halts the engine in the worst case. Methods to regenerate DPF are classified into two categories: high-temperature regeneration methods such as heat addition with an electric heater, afterburner, etc., and low-temperature regeneration methods or continuous regeneration methods such as CRT or NTP regeneration.

Reduction technologies for NO_x emissions include burning improvement techniques such as the control of fuel injection timing, lean-burn, and EGR. However, it is difficult to meet the more severe NO_x emission standards in the future only using burning improvement techniques. The establishment of low-cost and effective postprocess NO_x removal technologies is required. Because in general, 10% oxygen is contained in diesel engine emissions, a three-way catalyst used in the purification of gasoline engine emissions may not work for diesel emissions. Therefore, many NO_x reduction methods for diesel engine emissions have been proposed. At present, the most popular method is SCR that requires a high-performance reduction catalyst and infrastructure for the supply of urea or ammonia.

Fig. 3.5.13 shows the treatment equipment for diesel emissions proposed by Johnson Matthey Inc. The CRT system [36] consists of an oxidation catalyst and a DPF. The DPF is regenerated by NO_2 induced by the oxidation catalyst. The CRT is followed by the SCR catalyst and ammonia slip catalyst. The injected urea—water solution is decomposed to NH_3 at high temperature before the SCR catalyst. Using this system, HC, CO, PM, NO_x, and slipped NH_3 can be simultaneously removed. Although there are several problems associated with this system, the system has been already implemented to diesel automobiles, especially in Europe,

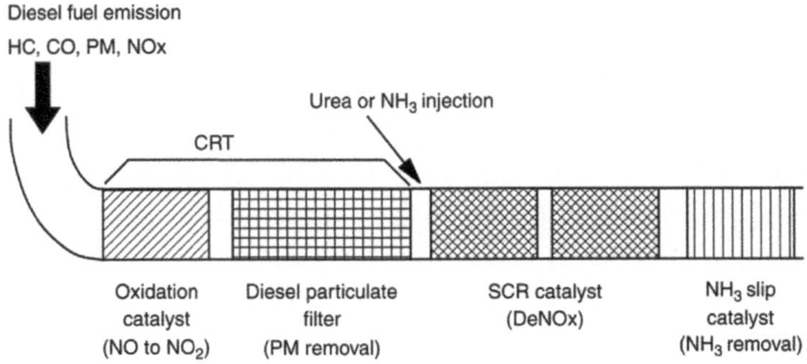

Figure 3.5.13 Final diesel engine treatment system. *CRT*, continuous regeneration trap; *PM*, particulate matter.

which has a high number of diesel automobiles. However, at lower exhaust gas temperature (< 200°C), nonthermal plasma treatment or ozone injection method is effectual. Ozone injection works well to induce NO_2. In the next, these methods are explained.

Nonthermal plasma regeneration methods

Various methods of DPF regeneration are illustrated in Fig. 3.5.14. The most typical regeneration process is shown in Fig. 3.5.14A. Soot removal or incineration is carried out under the oxygen-rich environment using an electric heater or high-temperature afterburner. However, because high temperatures around 600°C are needed for regeneration, melting or cracking of the DPF is possible.

A new regeneration method of NO_2 catalytic incineration system is shown in Fig. 3.5.14B. Incineration of soot begins at a low temperature of

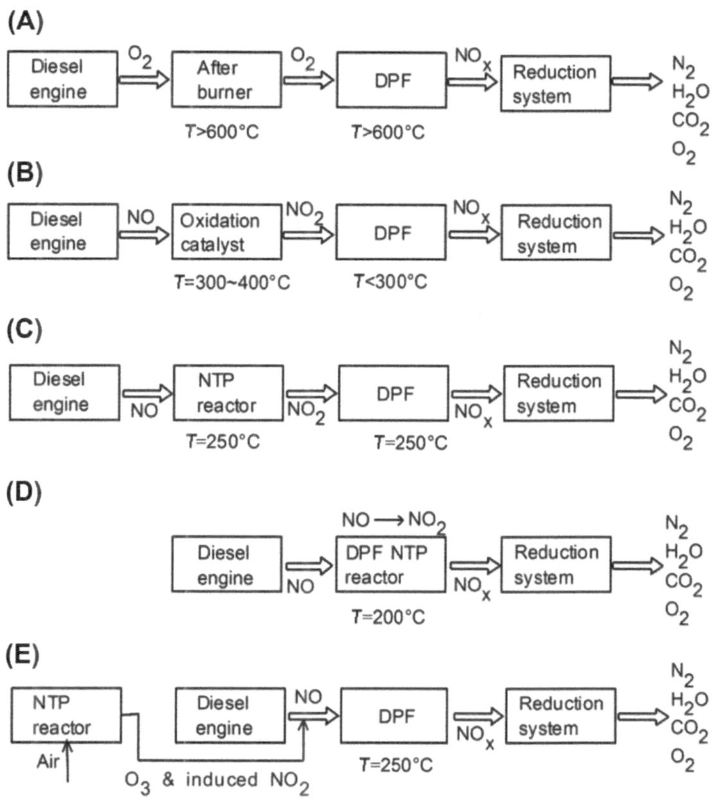

Figure 3.5.14 Methods of diesel particulate filter (DPF) regeneration. *NTP*, nonthermal plasma. *(Made based on Refs. [37] and [38].)*

200°C if a sufficient amount of NO_2 is available. Initially, NO from diesel engine exhaust is oxidized to NO_2 by oxidation catalysts using noble metals. Then, the soot deposited on the DPF reacts with generated NO_2 and is removed by incineration. However, a high temperature of 300–400°C is required to activate the oxidation catalysts. Moreover, the maximum efficiency of the oxidation catalyst is around 50%, and there are problems associated with the degradation of the oxidization catalyst under the influence of sulfur in light fuel.

More recently, a new regeneration method using NTP has been proposed as shown in Fig. 3.5.14C. The oxidation catalyst is replaced with a NTP reactor generally exhibiting the excellent performance of NO oxidation under low temperatures. Emitted NO_2 and radicals induced by the plasma reactor incinerate soot particles deposited inside the DPF at lower temperatures of less than 300°C. Because the oxidization catalyst is not used in this system, there are no problems associated with catalyst degradation.

A new compact NTP regeneration system in which DPF and a plasma reactor are unified is also being investigated as shown in Fig. 3.5.14D. In this system, the NTP induced in the honeycomb path of the DPF oxidizes NO to NO_2, and the induced NO_2 and radicals incinerate the captured soot particles simultaneously.

The ozone injection method shown in Fig. 3.5.14E is another powerful method of regeneration. The details of the method are explained in the next.

Regeneration principle

In PM removal using DPF and its regeneration as shown in Fig. 3.5.15, PM is firstly captured using a wall flow-type DPF. NTP-induced O_3 is secondly injected into the DPF before the exhaust gas passes through. Subsequently, the captured and accumulated PM is oxidized to gaseous CO_x (= CO + CO_2) by oxygen radicals (O), which are mainly produced by the thermal decomposition of O_3 and NO_2. Oxygen radicals and hydroxyl radicals (OH) contribute to the oxidation of the soot. The DPF is regenerated by this PM oxidation.

The chemical reactions of the above process are as follows [39]:

$$O_3 \rightarrow O_2 + O \quad (3.135)$$

$$NO + O \rightarrow NO_2 \quad (3.136)$$

$$NO_2 \rightarrow NO + O \quad (3.137)$$

$$C + 2O \rightarrow CO_2 \quad (3.138)$$

$$C + O \rightarrow CO \quad (3.139)$$

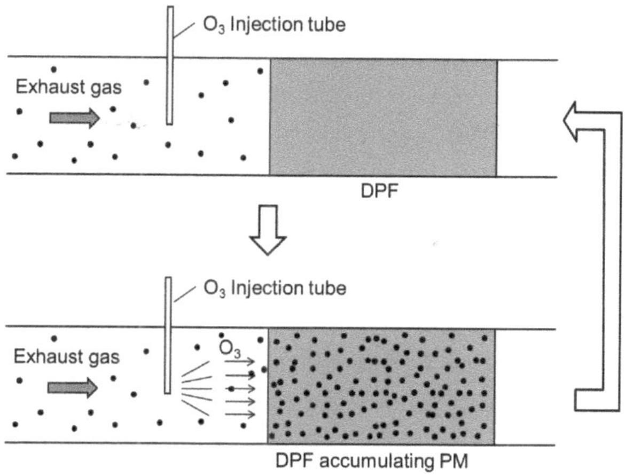

Figure 3.5.15 Process representation of plasma diesel particulate filter (DPF) regeneration. DPF regeneration is performed with nonthermal plasma–induced O_3 injection. *PM*, particulate matter.

$$SOF + lO \rightarrow mCO_2 + nH_2O \ (l,\ m,\ \text{and}\ n\ \text{are integers}) \quad (3.140)$$

$$2HC + 5O \rightarrow 2CO_2 + H_2O \quad (3.141)$$

$$2NO_2 + 2C \rightarrow N_2 + 2CO_2 \quad (3.142)$$

$$2NO_2 + 4C \rightarrow N_2 + 4CO \quad (3.143)$$

where HC and SOF mean hydrocarbon compounds and soluble organic fraction, respectively. It is noted that SOF is an abbreviation and does not express any chemical components. The typical reaction temperature is 200–300°C, which is relatively lower than that of the DPF regeneration by thermal oxidization (typically 600°C). NO_2 and O_3 primarily contribute to the oxidation of the PM. Because no catalyst is used in these reactions, the proposed method does not have the problem of catalyst degradation by sulfur.

3.5.4 Plasma-hybrid exhaust gas treatment

Basic concept

The atmospheric pressure NTP can be generated easily by applying an alternating current or a pulse high voltage between the electrodes placed on either side of an insulating barrier or a gap. Exhaust gas can be cleaned by flowing it between these electrodes; however, there are some confounding factors. For example, the energy efficiency (quantity of treatment per unit of plasma input energy in g/kWh) of the process must be increased. In general, if exhaust gas containing tens to hundreds mg/Nm^3 of PM and several hundred ppm of NO_x is discharged from an engine and flows between the electrodes, ordinary plasma treatment under proper conditions has an efficiency of no more than 1 g(PM)/kWh for PM, 10 g(NO_2)/kWh for NO reduction, and 50 g(NO_2)/kWh for the oxidation of NO to NO_2. However, for the treatment to be commercial, it is considered that treatment efficiency of 5 g(PM)/kWh for PM and that of 100 g(NO_2)/kWh for NO reduction are desired at least. Accordingly, in our process, named plasma-hybrid process the exhaust gas has not been treated directly. Instead of direct treatment, PM and NO_x are concentrated by accumulation on a filter or an adsorbent and subsequently, the concentrated polluting substances are treated with higher energy efficiency. This method is based on the chemical principle that higher concentration substances can be treated with higher efficiencies. This process has already been used for the treatment of VOCs and is also applicable to diesel exhaust gas.

Particulate matter treatment

Fig. 3.5.16 presents a schematic of our proposed system for the total cleaning of the diesel engine exhaust gas. The system employs plasma using the concepts outlined in the previous sections. The treatment of PM is described followed by the treatment of NO_x.

Diesel engine exhaust gas includes various types of PM. In addition to dry soot (carbon), commonly referred to as soot, it also includes a SOF, hydrocarbon compounds, and sulfates. As shown in Fig. 3.5.16, in our process, PM in the exhaust gas is firstly collected on a ceramic DPF. Secondly, O_3 produced by NTP is injected upstream from the DPF and cleaning by forming CO_x (= $CO + CO_2$), and water (H_2O) is achieved by O_3 and O radicals formed during the thermolysis of the resulting NO_2 according to reactions (3.135)−(3.143). These reactions present the

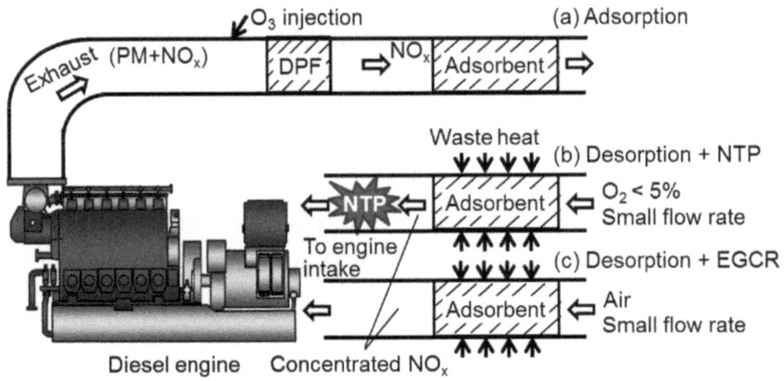

Figure 3.5.16 Schematic diagram of a plasma combined exhaust gas treatment system. *EGCR*, exhaust gas component recirculation. *NTP*, nonthermal plasma; *PM*, particulate matter.

simultaneous removal of dry soot by oxidation and NO_x by the reduction in plasma (plasma-assisted diesel particulate and NO_x reduction, PDPNR). These reactions are activated at low oxygen concentrations of < 5% [12].

In the case of diesel engines, the oxygen concentration in the exhaust gas is typically 11%, and in the case of a gas boiler, it is about 6.7%. However, to use NTP to form O_3 efficiently for PM oxidation reactions, a higher oxygen concentration is required. Therefore, a process in which O_3 is produced by an NTP ozonizer in the air with the oxygen concentration of 21% or in a pure oxygen atmosphere which can be provided by a pressure swing adsorption oxygen generator and injected into the exhaust gas clearly shows greater oxidizing efficiency than the generation of plasma and the formation of O_3 inside the exhaust gas. This process is termed the indirect plasma method, remote plasma method, or ozone injection method. In the process shown in Fig. 3.5.16, O_3 is injected upstream from the DPF. Compared with the direct plasma method, there is no need to introduce polluted exhaust gas into the plasma reactor; therefore, the electrodes are not soiled, which is advantageous for the durability of the reactor. However, compared with the direct plasma method, the simultaneous PDPNR of reactions (3.142) and (3.143) cannot be expected. The lifetime of N radicals, which contribute negatively to the reduction of NO_2 in reactions (3.142) and (3.143), is around a millisecond, and even if they are formed by NTP, they are almost completely removed by recombination by the time they enter the exhaust conduit. In contrast, the indirect plasma method has several advantages: When SO_2, a constituent of exhaust gas from burning heavy oil, oxidizes, it produces sulfate droplets and, as PM, these are difficult to collect in a DPF and to regenerate. However, with the indirect plasma method, the process of SO_2 oxidation is slower than that of NO, being essentially negligible. Consequently, SO_2 passes through the DPF as a gas, improving the DPF conditions. PM oxidation by plasma has been described and used together with the collection by DPF. The technology shows high energy efficiency in the order of 5–10 g(PM)/kWh. Therefore, it can be expected to mature enough for practical use soon.

NO_x treatment

Treating NO_x with plasma is more difficult than treating PM; however, the basic plasma reaction process is well understood. In general, when plasma is applied to NO_x at concentrations of several hundred ppm (over 90% NO) discharged from an incinerator, the N_2 and O_2 bonds, which are the main constituents of exhaust gas, are cleaved by high-speed electrons, and N and O radicals with unpaired electrons are formed. Thus, the reduction reaction (3.144) and the oxidation reaction (3.145) occur in parallel.

$$NO + N \rightarrow N_2 + O \tag{3.144}$$

$$NO + O\ (+H_2O) \rightarrow NO_2, N_2O_5, (HNO_3) \tag{3.145}$$

Data are obtained showing that at an oxygen concentration of $O_2 <$ 5%, reaction (3.144) is dominant, whereas at $O_2 >$ 5%, reaction (3.145) becomes significant. In addition, at $O_2 <$ 5%, some amount of NO is converted to N_2O, but the amount converted is not large (the conversion efficiency is only few percents). In addition, moisture in the air forms OH radicals; however, while some sources state that these oxidize NO, in our experience, when plasma is applied in an ordinary atmosphere or in the exhaust gas, the proportion of the oxidized NO is small and the oxidation reaction (3.145), which is driven by O radicals, is dominant. Recent experimental results [40] indicate that when NO and O react at substantially equimolar concentrations according to reaction (3.145), NO_2 is formed. Furthermore, when NO_x reacts with O in excess of equimolar quantities, oxidation to N_2O_5 and HNO_3 occurs, consistent with our results. Notably, the moisture concentration in marine diesel engine exhaust gas is very high. Because N_2O_5 and HNO_3 dissolve extremely readily in the moisture of the exhaust gas, care must be taken during measurements. The pollution treatment technology cannot be discussed based only on the oxidation reaction (3.145); the reduction reaction (3.144) must also be considered.

In Fig. 3.5.16, based on this background, a process is proposed in which (1) adsorption by an adsorbent (process (a) in the figure), (2) desorption by using heat from the engine (process (b) in the figure), NO_x concentration, and plasma treatment in a stream of low-concentration oxygen gas with O_2 < 5% at a low flow rate, and (3) NO_x treatment by repeated exhaust gas component recirculation (EGCR) by adsorption and desorption (process (c) in the figure) are carried out, as shown in the downstream of the PM treatment component. Furthermore, recently, experiments concerning the cleaning of marine diesel engines have been conducted.

As the NTP employed for PM and NO_x treatments, pulse corona plasma, barrier discharge plasma, or surface discharge plasma are also frequently used. It is preferable to use surface discharge plasma, which is cheap and has high energy conversion efficiency. The principles and an example of the plasma used for NO_x treatment (process (b) in Fig. 3.5.16)

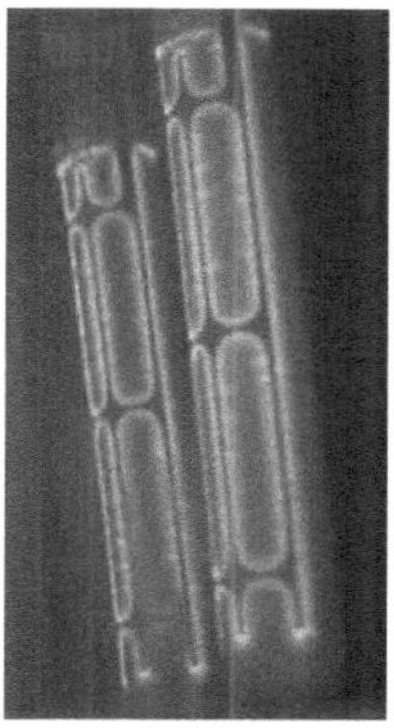

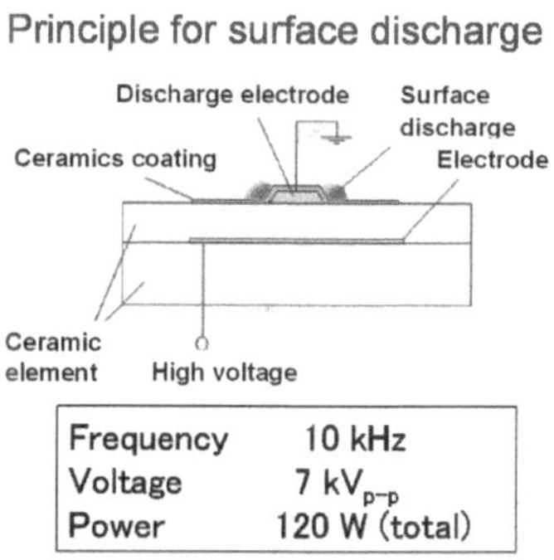

Figure 3.5.17 Generation of surface discharge–excited atmospheric pressure nonthermal plasma.

are presented in Fig. 3.5.17. By applying a bipolar pulse high voltage of 7 $kV_{p\text{-}p}$ (peak-to-peak voltage) and a frequency of 10 kHz between the surface discharge electrodes placed along the surface and high-voltage electrodes placed inside the ceramic element such as alumina, a plasma discharge is generated along the edges of the surface electrodes as shown in the photograph. Many of these rods are placed in the flow path and the target NO_x gas is passed through a narrow flow path to meet the plasma for cleaning.

Next, the treatment process (c), illustrated in Fig. 3.5.16, is described. To assist the reduction for cleaning by plasma, during the air-cooling step, an adsorbent is used to adsorb some of the NO_x that is not amenable to treatment by the plasma alone. However, some of the NO_x flows out of the absorbent. NO_x in the exhaust is concentrated, on the order of 2%; when reintroduced into the engine from the engine air intake, NO_x decomposes inside the engine due to the reducing action of hydrocarbons, hydrogen, and carbon monoxide. This step is termed EGCR and was discovered by us. Importantly, EGCR increases the energy efficiency of NO_x, increasing the cleaning of the exhaust by more than twofold. Total cleaning of marine diesel engine exhaust gas is carried out using the plasma treatment for the PM and NO_x cleaning processes. Examples of experiments carried out by the authors will be presented in Subsection 4.7.1.

The application of gas discharge plasmas has assumed an important place in many manufacturing processes. Examples and emerging applications, as well as the potential of plasma technologies, are discussed further in Ref. [41] with many applications.

References

[1] M. Sadakata, Taiki Kuriinka no Tame no Kagaku Kougaku (Chemical Engineering for Atmospheric Clean-up), Baifukan Publishing Co., 1999 (in Japanese).
[2] Chapter 3 on pp. 47−90 of Ref. [1].
[3] S. Katoh, T. Miydera, Taiki Tokuron (Special Lecture on Atmospheric Environment), Japan Environmental Management Association for Industry, 2009, pp. 20−43 (in Japanese).
[4] Y.B. Zeldovich, The oxidation of nitrogen in combustion explosions, Acta Physiochimica USSR 21 (4) (1946) 577−628.
[5] C.P. Fenimore, Formation of nitric oxide in premixed hydrocarbon flames, Proc. Combust. Inst. Symp. Int. Combust. 13 (1) (1971) 373−380.
[6] R.K. Lyon, The NH_3-NO-O_2 reaction, Int. J. Chem. Kinet. 8 (1976) 315−318.
[7] Y. Hamada, K. Aika, Bussitsu Kankyo Kagaku (Materials and Environmental Science), Textbook for School of Graduate Studies, the Open University of Japan, 2009, p. 166 (in Japanese).
[8] Y. Hayakawa, Y. Inoue, A. Takeyama, S. Kambara, Reaction mechanism of De-NO_x by activated ammonia generated by dielectric barrier discharge, Int. J. Plasma Environ. Sci. Technol. 10 (1) (2016) 20−23.
[9] G. Liu, P.-X. Gao, A review of NO_x storage/reduction catalysts: mechanism, materials and degradation studies, Catal. Sci. Technol. 1 (2011) 522−568.
[10] S. Roy, A. Baiker, NO_x storage−reduction catalysis: from mechanism and materials properties to storage−reduction performance, Chem. Rev. 109 (9) (2009) 4054−4091.
[11] R.D. Clayton, M.P. Harold, V. Balakotaiah, NO_x storage and reduction with H_2 on Pt/BaO/Al_2O_3 monolith: spatio-temporal resolution of product distribution, Appl. Catal. B Environ. 84 (3−4) (2008) 616−630.
[12] M. Okubo, T. Kuroki, K. Yoshida, T. Yamamoto, Single-stage simultaneous reduction of diesel particulate and NO_x using oxygen-lean nonthermal plasma application, IEEE Trans. Ind. Appl. 46 (6) (2010) 2143−2150.
[13] M. Zheng, G.T. Reader, J.G. Hawley, Diesel engine exhaust gas recirculation−a review on advanced and novel concepts, Energy Convers. Manag. 45 (6) (2004) 883−900.
[14] Kankyo Hozen Taisaku Kenkyukai, eds., Taiki Osen Taisaku no Kiso Chishiki, (Fundamentals and Knowledge for Atmospheric Aftertreatment System), 2001, Japan Environmental Management Association for Industry, pp. 80−85, (in Japanese).
[15] H. Yamamoto, T. Kuroki, H. Fujishima, M. Okubo, Pilot-scale exhaust gas treatment for a glass manufacturing system using a plasma combined wet chemical process, Mech. Eng. J. Jpn. Soc. Mech. Eng. 3 (1) (2016), 15-00549, total 10 pages.
[16] S. Aoki, R. Suzuki, Electron-beam flue gas treatment system, IEEJ Trans. Fundam. Mater. 114 (5) (1994) 349−355.
[17] J.S. Clements, A. Mizuno, W.C. Finney, R.H. Davis, Combined removal of SO_2, NO_x, and fly ash from simulated flue gas using pulsed streamer corona, IEEE Trans. Ind. Appl. 25 (1) (1989) 62−69.
[18] S. Li, Y. Huang, F. Wang, J. Liu, F. Feng, X. Shen, Q. Zheng, Z. Liu, L. Wang, K. Yan, Fundamentals and environmental applications of non-thermal plasmas: multi-pollutants emission control from coal-fired flue gas, Plasma Chem. Plasma Process. 34 (3) (2014) 579−603.
[19] M. Nonokawa, T. Furui, S. Noshiro, PM reduction technology using the Cerallec® system ceramic filter for marine diesel engines, J. Japan Inst. Mar. Eng. 48 (4) (2013) 505−509 (in Japanese).

[20] K. Saitoh, M. Shima, Y. Yoda, R. Nakatsubo, T. Hiraki, Composition characteristics of size-resolved airborne particles in Himeji city, NMCC Ann. Rep. 16 (2009) 163–165 (in Japanese).
[21] C.D. Cooper, F.C. Alley, Air Pollution Control (A Design Approach), second ed., Waveland Press, Inc., 1994 (Chapter 5).
[22] H.J. White, Industrial Electrostatic Precipitation, Addison-Wesley, Reading, MA, 1963.
[23] Chapter 4 on pp. 91–145 of Ref. [1].
[24] W.C. Hinds, Aerosol Technology, John Wiley & Sons, Inc, USA, 1982. Chapters 3 and 5.
[25] Advanced physicochemical treatment technologies, in: L.K. Wang, Y.-T. Hung, N.K. Shammas (Eds.), Handbook of Environmental Engineering, vol. 5, The Humana Press Inc., USA, 2007, Chapter 4, Nonthermal plasma technology written by T. Yamamoto and M. Okubo.
[26] J. Adler, Ceramic diesel particulate filters, Int. J. Appl. Ceram. Technol. 2 (6) (2005) 429–439.
[27] Johnson Matthey's website, http://www.jmj.co.jp/diesel/crt.html.
[28] K. Naito, S. Sembayashi, Development of a new type diesel particulate filter (DPF), Nissin Electr. Rev. 57 (1) (2012) 37–42 (in Japanese).
[29] S. Okamoto, Y. Ichikawa, Kankyougaku Gairon (Textbook of Environmental Study), second ed., Sangyo Tosho Co., 2005 (in Japanese).
[30] F.F. Chen, Introduction to Plasma Physics, Plenum Press, New York, NY, 1974.
[31] M. Mitchner, C.H. Kruger, Partially Ionized Gas, John Wiley and Sons, New York, NY, 1973.
[32] J.E. Drummond, Plasma Physics, Dover Books on Physics, Dover Publications, Inc, Mineola, New York, 2013.
[33] J.S. Chang, P.A. Lawless, T. Yamamoto, Corona discharge processes, IEEE Trans. Plasma Sci. 19 (6) (1991) 1152–1166.
[34] M. Kuzumoto, Extremely narrow discharge gap ozone generator, J. Plasma Fusion Res. 74 (10) (1998) 1144–1150 (in Japanese).
[35] Masuda Research Inc. website, https://masuda-research.co.jp/general.html#GENERAL; https://masuda-research.co.jp/inquiry/OZS-70HP.pdf.
[36] B.J. Cooper, The catalytic control of motor vehicle emissions, Preprint of Commemorative Lecture at the Twenty-Third Honda Prize Awarding Ceremony, Tokyo, November 15, 2002.
[37] M. Okubo, T. Miyashita, T. Kuroki, S. Miwa, T. Yamamoto, Regeneration of diesel particulate filter using nonthermal plasma without catalyst, IEEE Trans. Ind. Appl. 40 (6) (2004) 1451–1458.
[38] M. Okubo, T. Kuroki, Y. Miyairi, T. Yamamoto, Low-temperature soot incineration of diesel particulate filter using remote nonthermal plasma induced by a pulsed barrier discharge, IEEE Trans. Ind. Appl. 40 (6) (2004) 1504–1512.
[39] T. Kuwahara, S. Nishii, T. Kuroki, M. Okubo, Complete regeneration characteristics of diesel particulate filter using ozone injection, Appl. Energy 111 (2013) 652–656.
[40] K. Skalska, J.S. Miller, S. Ledakowicz, Intensification of NO_x absorption process by means of ozone injection into exhaust gas stream, Chem. Eng. Process 61 (2012) 69–74.
[41] K.-D. Weltmann, J.F. Kolb, M. Holub, D. Uhrlandt, M. Š imek, K.K. Ostrikov, S. Hamaguchi, U. Cvelbar, M. Černák, B. Locke, A. Fridman, P. Favia, K. Becker, The future for plasma science and technology, Plasma Process. Polym. 16 (2019) e1800118.

CHAPTER 4

Operation examples of emission control systems

Contents

New Technologies for Emission Control in Marine Diesel Engines
ISBN 978-0-12-812307-2
https://doi.org/10.1016/B978-0-12-812307-2.00004-3

4.1 Selective catalytic reduction

4.1.1 Laboratory-scale selective catalytic reduction

In addition to conventional selective catalytic reduction (SCR) technology, several techniques have been proposed and studied to enhance the conversion efficiency or to implement the technology at a low temperature. An example is a study of the application of SCR technology at a low temperature [1]. Among many SCR catalysts, those employed in a typical commercial SCR process are V_2O_5/TiO_2 mixed with WO_3 or MoO_3. As explained in Section 3.2, the following is one of the general SCR reactions:

$$4NO + 4NH_3 + O_2 \rightarrow 4N_2 + 6H_2O \tag{4.1}$$

The advantage of the vanadium (V)-based catalyst is that it is highly active and resistant to SO_x, which is a component of exhaust gas. However, as a disadvantage, its utilization is limited to the temperature range of 300−400°C, and N_2O is also formed. The increase in the fuel efficiency of marine diesel engines has resulted in a decrease in the temperature of the exhaust gas. Therefore, with the view of actual future utilization and installation, an alternative method for SO_2-resistant SCR catalysts that are active in the range of lower temperature less than 200°C are greatly desired.

To achieve NO_x reductions using SCR catalysts that are active in the range of 150−160°C, plasma-enhanced SCR techniques have been proposed and studied [2]. In nonthermal plasma (NTP) such as electrical discharges in the atmosphere, the following main reactions related to NO_x take place:

$$e + O_2 \rightarrow e + O + O \tag{4.2}$$

$$O + NO + M \rightarrow NO_2 + M \quad (4.3)$$

where M indicates the third body. On the other hand, among NO_x reductions over an ammonia SCR catalyst, it is known that the following reaction is most efficient:

$$2NH_3 + NO + NO_2 \rightarrow 2N_2 + 3H_2O \quad (4.4)$$

In this reaction, NO has the same number of moles as NO_2. However, the reaction cannot proceed at a low temperature of 150°C. Reactions (4.2) and (4.3) induced by NTP can assist the NO_x reduction reaction (4.4) over an ammonia SCR catalyst (zeolite type) at a low temperature of 150–160°C. Fig. 4.1.1 shows a schematic of catalyzed pellet packed-bed plasma reactor for NO_x reduction at a lower temperature. Firstly, experiments are carried out using model gas, that is, a dry air including 200 ppm of NO_x (NO = 100 ppm and NO_2 = 100 ppm) and 200 ppm of NH_3 ($NO_x/NH_3 = 1$). The model gas at the flow rate of 2 L/min is treated by the reactor, which is energized by a pulse high voltage power supply. When the high voltage is applied between the stainless steel rod and the aluminum foil, NTP is induced among the catalyzed pellets, and they are activated to realize NO_x reduction at a lower temperature. In the case of catalyzed pellet with 30 mL, NO_x reduction efficiency is 70% at a temperature of 110°C. Secondly, the experiments are carried out with a diesel engine (six-cylinder, displacement of 7.7 L). The concentration of NH_3 is 10 volume%. The discharge power is 1.9 W (19 $kV_{p\text{-}p}$, 5 kHz). The exhaust gas temperature is 140°C at the inlet of the NTP reactor that contains the catalyst. As a result, it is demonstrated that the NO_x reduction efficiency is 72% without NTP application and 88% with NTP application.

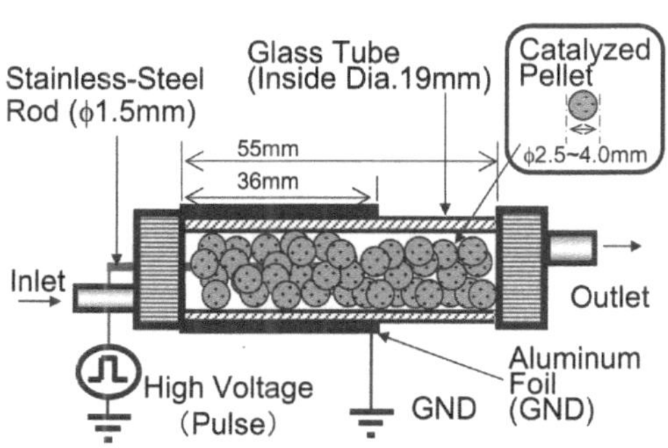

Figure 4.1.1 Schematic diagram of packed bed plasma reactor with catalyst pellets for NO_x reduction [2].

Plasma catalysis for environmental treatment and energy applications are reviewed in a paper [3], for example, ammonia synthesis for Hydrogen Society is discussed.

4.1.2 Pilot-scale selective catalytic reduction

SCR uses ammonia as a reducing agent. Ammonia is introduced into the exhaust stream from a combustor. This is a fundamental SCR process. It reduces the levels of NO_x within a catalyst system. Ammonia is a hazardous substance that requires careful management. Instead of the direct utilization ammonia, urea is used widely in a system known as the urea-SCR system. NH_3 generation from urea is explained in Chapter 3. An aqueous solution of urea for use in the urea-SCR system is called diesel exhaust fluid, which contains 32.5% of urea and 67.5% of deionized water. Although SCR technology using ammonia has been studied and developed for a long time and has been applied in thermal power generation plants since the 1970s, the urea-SCR technology has been subsequently implemented in automobiles and later in commercial diesel trucks since 2004 [4]. The motivation for applying the technology is to satisfy the regulations, which have become more stringent in each field. In the maritime field, the regulations are known as the International Maritime Organization (IMO) emission standards and are referred to as Tier I, II, and III standards on NO_x, as explained in Section 2.2. These regulations restrict the mass of NO_x emitted per engine output energy for a given engine rotating speed. This stringent requirement cannot be satisfied merely by an improvement of the combustion in diesel engines, and an aftertreatment technology is needed. In this regard, the SCR system has been developed as a leading technology.

An SCR system developed by Daihatsu Diesel MFG. Co., Ltd., Japan, is an example of pilot- or full-scale commercial SCR technology for marine diesel engines [5]. It has been developed to meet the requirements for IMO NO_x Tier III. The actual operations specified for entering and leaving the Emission Control Area (ECA) are as follows:

1. Fuel is changed from high- to low-sulfur content.
2. SCR operation is turned on before entering ECA.
3. SCR operation is turned off after leaving from ECA.
4. Fuel is changed from low- to high-sulfur content.

Application of the technology to marine diesel engines requires us to pay attention to the degradation of the SCR catalyst by sulfur such as by the SO_x in the exhaust gas. At temperatures lower than 300°C, ammonia (NH_3) reacts with sulfur oxide (SO_x). As a result, aerosols of ammonium hydrogen sulfate (NH_4HSO_4) and ammonium sulfate ($(NH_4)_2SO_4$) are generated. Both are a molten salt in the liquid state at the abovementioned temperature and cause the performance of the catalyst to decay. In the SCR system, this problem is addressed by installing a bypass flow path to overcome the degradation of the SCR catalyst. In the ECA, the degradation of the SCR catalyst is not an issue if low-sulfur fuel is utilized. In the exterior of the ECA, i.e., when high-sulfur fuel is used, the exhaust gas should flow out without making contact with the SCR catalyst.

Fig. 4.1.2 shows a schematic of the SCR system in (A) vertical and (B) horizontal installations. The bypass flow path and the SCR reactor are included in an all-in-one design to make the SCR system compact in consideration of the limited room on ships. The SCR unit can be installed

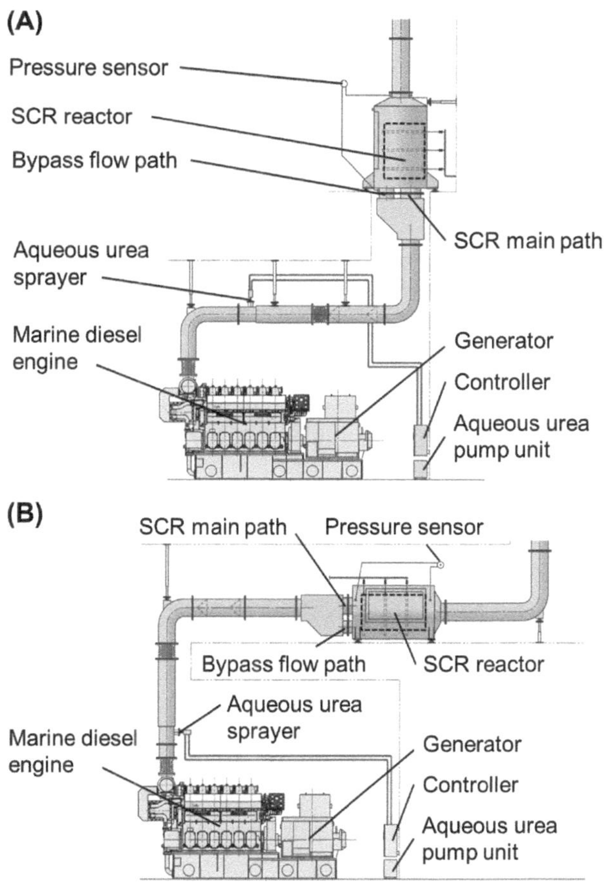

Figure 4.1.2 Schematic drawing of Daihatsu selective catalytic reduction system in (A) vertical and (B) horizontal installations. *(Made based on Ref. [5].)*

vertically or horizontally depending on the available space. The switching of a flow path between SCR main path and bypass flow path is done by electronic control. The relationship between the concentration of NO_x in the emission and the injection flow rate of the urea solution was determined in advance. The concentration of NO_x in the emission is inversely proportional to the injection flow rate of the urea solution. The range of the injection flow rate is adjusted such that it fulfills the requirements of IMO Tier III. On the basis of the data, the SCR system is controlled by a program that considers the relationship between the engine load and the injection flow rate of the urea solution. That is, the SCR system can maintain the emitted NO_x concentration at a low level by rapidly adjusting the injection rate of the urea solution against changes in the engine load to comply with the requirements of IMO Tier III. In addition to the stationary test, the performance of the system was tested on the 8600-TEU (twenty-foot equivalent unit) class container ship, whose name is Hanoi Bridge, (Operator: Kawasaki Kisen Kaisha, Ltd., Japan, Builder: Japan Marine United Corporation, Japan) during oceanic operation. The target engine was a diesel generator engine (8DC-32, Daihatsu Diesel MFG. Co., Ltd.). Exhaust gas from four diesel engine generators was treated. The main specifications of the engine were as follows: output power of 3000 kW, four-cycle, eight cylinders with a cylinder bore of 320 mm and a stroke of 400 mm, and a constant rotation rate of 720/750 rpm. The test confirmed that the degradation of the SCR catalyst did not occur during repeated operations.

Another pilot-scale example of SCR is reviewed. A marine SCR system developed by Hitachi Zosen Corporation, Osaka, Japan, is another example of pilot- or full-scale SCR technology for marine diesel engines [6,7]. Fig. 4.1.3 shows an illustration and photographic image of a marine SCR system. The system comprises a two-cycle marine diesel engine (6S46MC-C7, Hitachi-MAN B&W), turbocharger, exhaust receiver, vaporizer-containing urea injector and mixer, and SCR reactor containing catalyst blocks and soot blowers. The main specifications of the engine are as follows: maximum output power of 6780 kW, two-cycle, six cylinders with a cylinder bore of 460 mm and a stroke of 1932 mm, and a rotation rate of 111 rpm. In this system, the SCR unit is installed upstream of the turbine of the turbocharger. The directions of exhaust gas flows are shown with two types of arrows in the cases with SCR and non-SCR operations. Fig. 4.1.4 shows the schematic diagram of the marine SCR system, which corresponds to Fig. 4.1.3. The exhaust gas flows into the

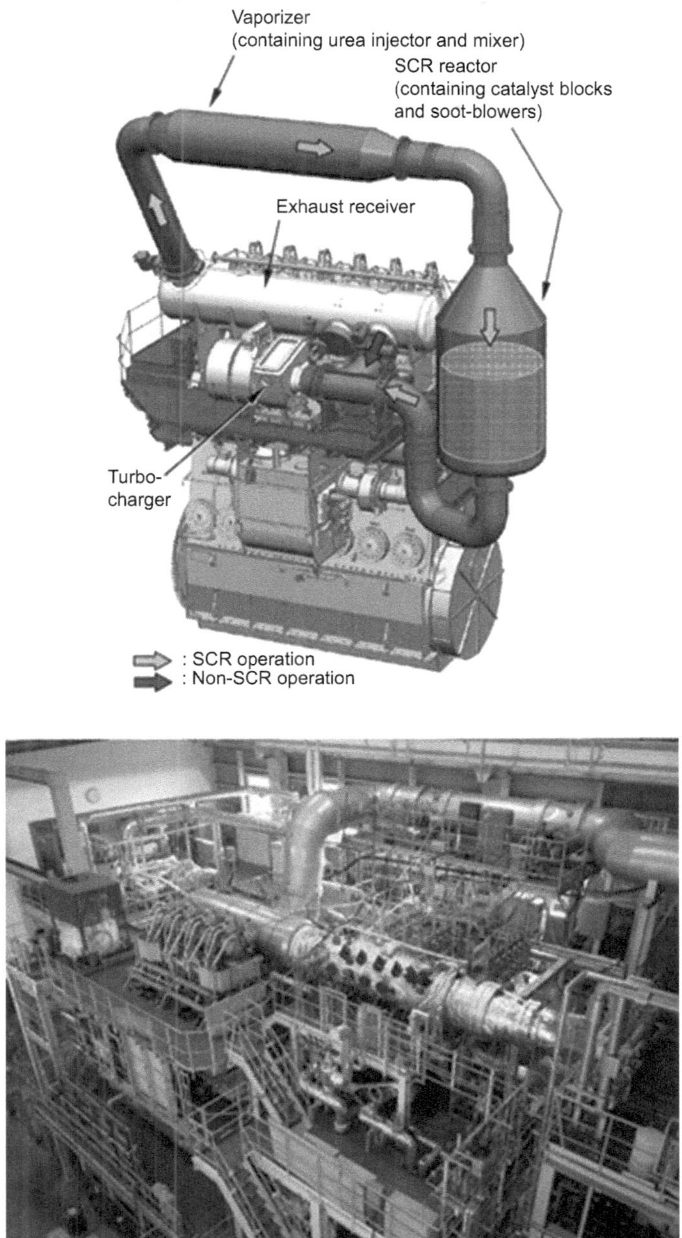

Figure 4.1.3 Schematic drawing and photographic image of the marine SCR system [7]. *SCR*, selective catalytic reduction.

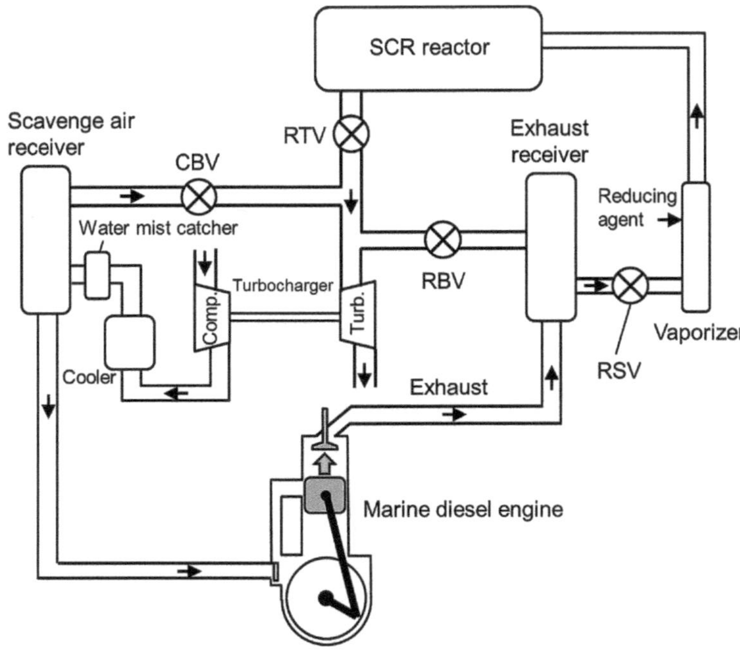

Figure 4.1.4 Schematic diagram of the marine SCR system [7]. *CBV*, cylinder bypass valve; *RBV*, reactor bypass valve; *RSV*, reactor sealing valve; *RTV*, reactor throttle valve; *SCR*, selective catalytic reduction.

receiver followed by the vaporizer, where reducing agent or urea solution is injected into the exhaust gas. The gas mixture flows into the SCR reactor in which the SCR catalyst is placed, and NO_x reduction proceeds. The treated gas is supplied to the turbine and flows out of the system. The part of the intake air from the compressor of the turbocharger is introduced to the turbine using the air scavenge air receiver. This type of SCR, which is known as a high-pressure SCR, can be used to easily obtain exhaust gas at high pressures and at temperatures higher than 300°C because an SCR reactor is installed on the upstream of the turbine. Therefore, the active reduction of NO_x is achieved, and the degradation of the SCR catalyst can be prevented. Because NO_x is actively and rapidly reduced with the SCR catalyst, the SCR unit can be compact having a honeycomb structure. In general, the disadvantage of a high-pressure SCR is that the turbine response is not satisfactory because the vaporizer and SCR reactor are installed between the engine exhaust and the turbine. Therefore, the high-pressure SCR system has not been practically implemented on ships. The present marine SCR system overcomes this issue by using automatic control with three valves, i.e., the reactor sealing valve (RSV), reactor throttle valve (RTV), and reactor bypass valve (RBV).

A stationary test was carried out. In the operation without SCR treatment, the amounts of emitted NO_x were approximately 18 g/kWh, 17 g/kWh, 15 g/kWh, and 12 g/kWh at an engine load of 25%, 50%, 75%, and 100%, respectively. More specifically, the weighted average for all the engine loads was 14.6 g/kWh, which fulfills the IMO Tier I (17.0 g/kWh) requirement. In the calculation of the weighted average, weighting factors of 0.15, 0.15, 0.5, and 0.2 are applied to the amounts of emitted NO_x for 25%, 50%, 75%, and 100%, respectively. These are weighting factors in the test cycle called E3 test cycle. In operations with SCR aftertreatment, the resulting weighted average for all engine loads was 2.8 g/kWh, which fulfills the IMO Tier III (3.4 g/kWh) requirement. In addition to the stationary test, during oceanic operations, the performance of the system was tested on a type 38,000-DWT (deadweight tonnage) general cargo ship. As a result, even at low engine loads, the necessary temperature of 300°C was maintained by controlling the system with a cylinder bypass valve (CBV). This is because, when the CBV is open to reduce the air supplied from the scavenge air receiver to the marine diesel engine, the temperature of exhaust gas from the engineer becomes higher. During SCR operations, the pressure and turbine rotation of the turbocharger was stable owing to the fast response to changes in the engine load using the valve control. During operation with SCR treatment, for all engine loads, the resulting weighted average for all engine loads was 3.1 g/kWh, which is better than the value of the stationary test and fulfills IMO Tier III requirements.

4.2 SO_x scrubber

4.2.1 SO_x emission and MARPOL convention

The release of marine engine exhaust gas into the atmosphere is needed to be lowered especially to suppress the emission of SO_x from marine diesel engines. In this regard, the MARPOL convention Annex VI, Regulation 14 was issued to regulate the sulfur concentration in marine fuels, and the regulation regularly becomes more stringent. This regulation requires marine fuel including a sulfur concentration of less than 0.1 mass% to be used after January 1, 2015 in the ECA of the North Sea, the Baltic Sea in Europe, and the seashore areas of North America, etc. On the other hand, in sea areas other than those in the ECA, marine fuels including a sulfur concentration less than 0.5% must be used after January 1, 2020. The detail of the regulation is explained in Section 2.3. Because the ECA area may be extended hereafter, stringent regional regulations may be introduced by many countries [8].

The MARPOL convention, Annex VI Regulation 14 determines that an exhaust gas cleaning system (EGCS), which can reduce SO_x emitted from combustion engines, may be used instead of low sulfur fuel. Compared with ordinary marine fuel oil, low-sulfur fuel is more expensive, and the ability of the fuel supplier to maintain the supply in view of the future increase in the demand is unclear. Therefore, the EGCS has attracted much attention because it allows the continued use of the usual and sufficiently supplied high-sulfur marine diesel fuel. Requests for the use of the EGCS have increased especially for ships entering the ECA, and rapid development is observed. The fundamentals and principles of the SO_x removal scrubber are explained in Section 3.3. In the following subsections, the results of the application of the scrubber to real marine diesel engines are described.

4.2.2 SO_x scrubber system for marine diesel engines

A SO_x scrubber demonstration plant is constructed by Mitsubishi Heavy Industries, Ltd. collaborated with Mitsubishi Kakoki Kaisha, Ltd. The plant is explained here based on Ref. [8]. The SO_x scrubber is an EGCS that can remove SO_x from emissions of marine diesel engines, and large demand exists. A demonstration plant including a SO_x scrubber was constructed in the Nagasaki factory, Japan, of Mitsubishi Heavy Industries for a two-stroke marine diesel single-cylinder engine bench (regulated power = 762 kW). Many kinds of prototype tests were performed, and data on the process of decomposition of sulfur oxide (De-SO_x) for marine diesel emission were stored. As a result, De-SO_x exceeding 98% was confirmed. This performance can fully satisfy the MARPOL convention regulation for SO_x. The SO_x scrubber system is developed based on the EGCS guideline of IMO.

4.2.3 Overview of SO_x scrubber system

The present EGCS has two gas cleaning modes as following [8]:

Freshwater closed-cycle mode: In the mode pure cleaning water containing sodium hydroxide (NaOH) is used for the neutralization of acid in the exhaust gas, as shown in Fig. 4.2.1. The exhaust gas from diesel engines is introduced from the exhaust pipe of the engine to the scrubber tower. Scrubber seawater is pumped up with pumps from the sea and introduced to the scrubber

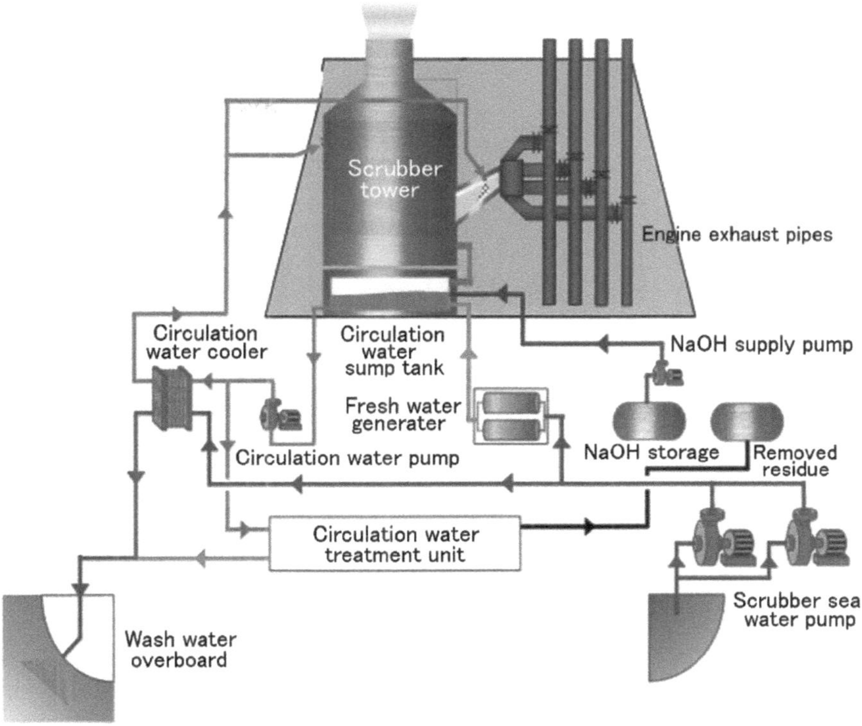

Figure 4.2.1 Freshwater-closed-cycle mode in the SO_x scrubber system [8].

tower and injected at the inlet pipe and the tower. The exhaust gas contacts with water and water-soluble components such as SO_2, SO_3, and NO_2 are treated or absorbed in the solution. On the other hand, NaOH is supplied to the scrubber tower (liquid NaOH storage at the lower region of the tower) and a mixture of seawater and NaOH solution is circulating with a pump. During the circulation, the properties of circulating solution such as pH are monitored. Before the solution is drained to the sea, it passes through the wastewater treatment apparatus and is cleaned to satisfy the regulation.

Seawater open-cycle mode: In the mode, seawater is taken from the sea and showered on the exhaust gas, as shown in Fig. 4.2.2. In the mode, the exhaust gas from diesel engines is introduced from the exhaust pipe of the engine to the scrubber tower. Scrubber seawater is pumped up with pumps from the sea and introduced to the scrubber tower and injected at the inlet pipe and the tower. The exhaust gas contacts with the seawater, and water-soluble components such as SO_2, SO_3, and NO_2 are absorbed into the solution. The wastewater is drained to the sea after pH control.

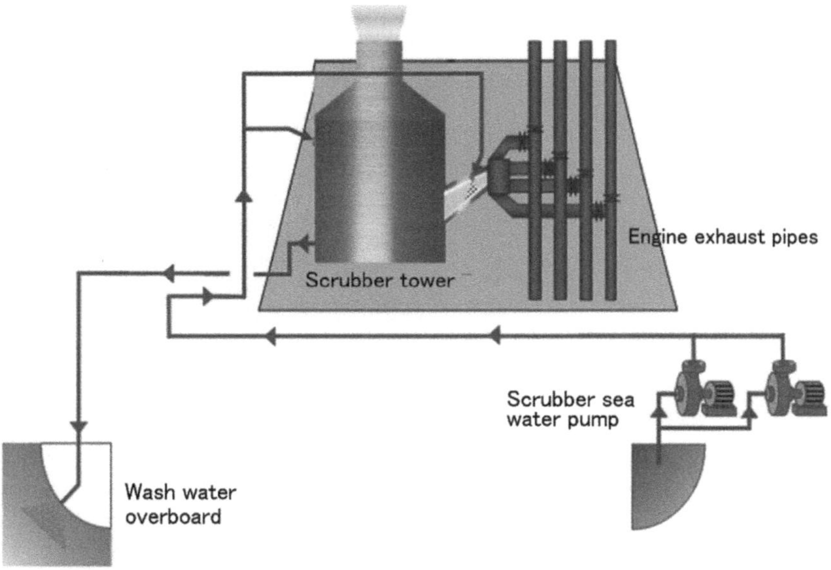

Figure 4.2.2 Seawater open-cycle mode in the SO_x scrubber system [8].

The constructed system is a hybrid type and can be used in either of the modes. It is possible to use De-SO_x for marine diesel engine emission for ships entering various areas such as the open sea, rivers, and ports.

4.2.4 Characteristics of SO_x scrubber system

All the components of the SO_x scrubber system are entirely controlled by the SO_x scrubber operation panel located in the engine control room. Further, the start-stop operation of the system and change over to washing mode are automatically performed by using a simple touch-panel interface. No complicated valve operations are required. The main system, including the circulation water pump, heat exchanger, wastewater treatment equipment, NaOH tank, and residual tank, can be supplied as an ISO-based container module. The regular container module is designed to utilize the limited space in ships effectively. Further, the system can be manufactured in a short period. In some kinds of ships using the container module system, the main components of the system can be located outside the ship's body. Fig. 4.2.3 shows an example of the divergent layout of the components inside the ship. It is also possible to retrofit the system to a moving ship.

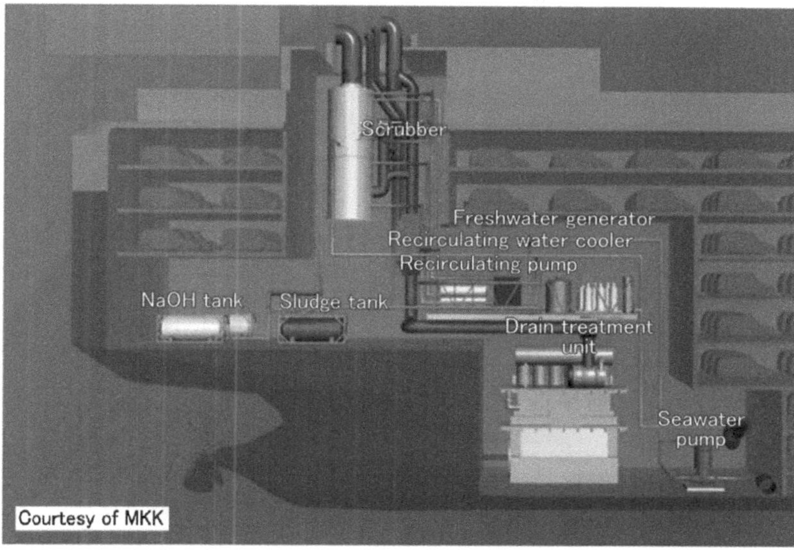

Figure 4.2.3 SO_x scrubber system installed in a ship [8].

4.2.5 Application of SO_x scrubber system to real ship

This SO_x scrubber system was mounted on a large ship of an automobile carrier in which 7500 automobiles can be accommodated, and onboard sea trials were performed. In the SO_x scrubber system of the ship, the exhaust pipes from the main diesel engines and three sub power diesel engines are connected to the scrubber tower located inside the engine casing. The exhaust gas emitted from all the 15-MW class engines can be simultaneously cleaned. The De-SO_x performance of the SO_x scrubber system mounted in the ship was tested during a test drive at sea. The measured relations between the efficiency of SO_x emission reduction and the ratio of the cleaning liquid flow rate to the exhaust gas one or liquid/gas ratio are plotted in Fig. 4.2.4. The graph shows that the SO_x in the exhaust gas can be removed to the satisfactory level of De-SO_x efficiency corresponding to 0.1% sulfur in fuel oil, which is demanded in the EGCS guideline in both the pure water-circulating mode and seawater single-pass mode. For both sea trials and prototype tests, satisfactory results are obtained. Therefore, it is confirmed that the present system has high De-SO_x performance for marine diesel exhaust gas.

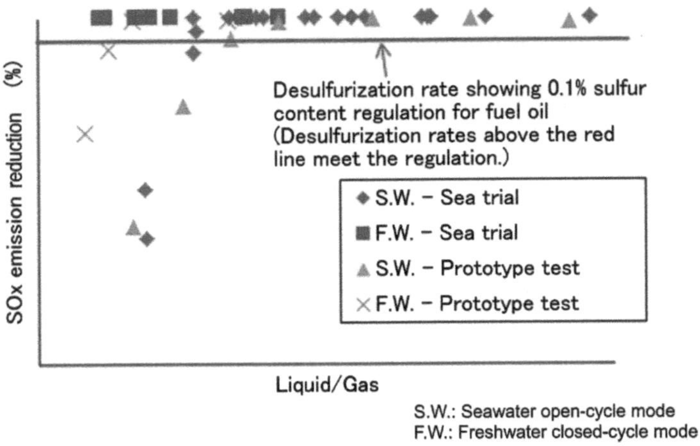

Figure 4.2.4 Measured relation between De-SO_x efficiency and the liquid-to-gas ratio of the cleaning solution [8].

4.2.6 Sodium hydroxide preparation using electrolyzing seawater

In EGCSs applied to the desulfurization of marine diesel engines, the supply of sodium hydroxide (NaOH) is an important problem. Because NaOH is a relatively expensive chemical and a strong alkali, careful handling is required when mounting it in large quantities onboard a ship. On the other hand, because seawater contains sodium ions, if a sufficient amount of NaOH through the electrolysis of seawater can be obtained, it would be very useful for the EGCS. Although the electrolysis of seawater has been studied for a long time, a pioneering study by Nishida et al. [9] is aimed at the desulfurization of marine diesel engines.

Fig. 4.2.5 shows a schematic of seawater electrolysis. As a weak alkali, seawater is an electrolyte solution, the chemistry of which is dominated by the presence of many ions including Na^+, Cl^-, and SO_4^{2-}. By applying a low voltage of several volts, ions are separated, and cations and anions move toward the respective negative and positive electrodes. Both electrode regions are separated using a membrane such as a polytetrafluoroethylene (PTFE) filter. Sodium ions, Na^+, gathering close to the negative electrode, react with OH^- ions given by water, and thus NaOH is produced in the negative electrode region. The electrolysis reactions can be described as follows:

$$\text{At a positive electrode: } Cl^- \rightarrow Cl_2/2 + e^- \tag{4.5}$$

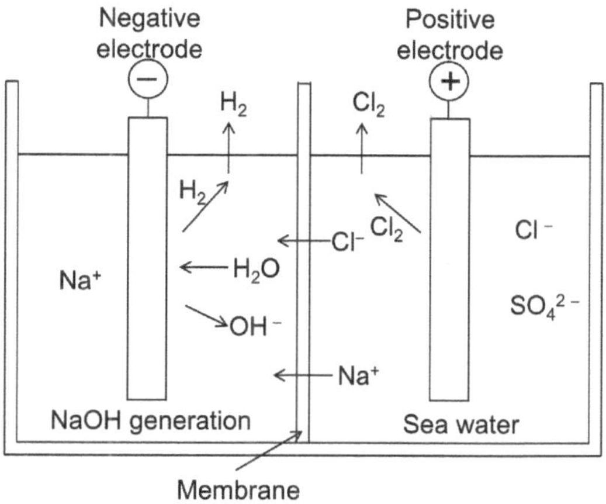

Figure 4.2.5 Schematic of seawater electrolysis [9].

$$\text{At a negative electrode: } H_2O + e^- \rightarrow H_2/2 + OH^- \quad (4.6)$$

$$Na^+ + OH^- \rightarrow NaOH \quad (4.7)$$

Consequently, an alkali solution of approximately pH = 10 or higher is created with H_2 and Cl_2 gas generation. The theoretical cell voltage is around 2.2 V, and it will be necessary to use 1.474 kWh of energy to obtain 1 m^3 of 100% NaOH. Further improvement on the energy efficiency for NaOH generation is necessary for practical application of seawater electrolysis.

4.3 Electrostatic precipitator

4.3.1 SO_x and particulate matter simultaneous removal system for marine diesel engines

SO_x and particulate matter (PM) simultaneous removal system for marine diesel engines using electrostatic precipitator (EP) is explained in this subsection [10]. The MARPOL convention Annex VI, as explained in Appendix determines the schedule of SO_x and PM cleanup for marine diesel engines. The substitution of high sulfur content fuel oil (heavy fuel oil [HFO] or high-sulfur fuel oil [HSFO]) by low sulfur content fuel oil (marine gas oil [MGO]) for use as marine oil fuel makes it possible to simultaneously realize both SO_x and PM emission controls, which is strongly related to the SO_x concentration in the exhaust. This can lower the environmental impact of marine diesel engines on the atmospheric environment. In Fig. 2.3.2 in Chapter 2, after 2020, sulfur of less than 0.1% fuel must be used in ECA seas and sulfur of less than 0.5% fuel must be used in general sea, although EGCS for SO_x treatment system having the equivalent performance is permitted.

Because compared with HFO the price of MGO is higher, the economic impact should be considered. Enforcement of the use of MGO (sulfur $< 0.1\%$) in the marine transportation industry in ECA would increase the risk faced by enterprises due to the fuel cost. Oil refinery companies would have to make a considerable capital investment to counteract the increased demand for MGO, which would cause the CO_2 emissions of oil refinery factories to increase. For the marine manufacturing industries, MGO does not have superior lubrication characteristics; thus, an improvement in lubricant technologies for marine diesel engines would be required.

These requirements for marine diesel engines can be mitigated by considering the option of SO_x and PM regulation. HFO can be used in conjunction with shipping EGCS inside the ship, which delivers almost the same performance on SO_x emission as the usage of MGO. A guideline for the installation of the EGCS on ships was issued by the IMO. Further, the Europen Union (EU) instruction becomes valid, and the regulations were valid in 2010; therefore, many kinds of EGCS apparatus are being developed, and research has become active especially in EU countries. Most EGCSs use a wet chemical scrubber as the main apparatus. Treatment of a large amount of exhaust gas emitted from marine diesel engines requires the size of the apparatus to become very large. Especially for newly manufactured ships, NO_x treatment should also be considered, as well as SO_x and PM treatment. Therefore, EGCSs should preferably be designed with a small size, and their operating cost should be reduced. Based on this state of the marine engine field, a group of Fuji Electric Co. developed an EGCS capable of the simultaneous removal of SO_x and PM in 2010 [10].

Fig. 4.3.1 shows a concept figure of a developed EGCS. One of its characteristics is the EP followed by the seawater scrubber for SO_x removal. The volume of the waste solution of the scrubber can be reduced to avoid the use of an excessively large wastewater treatment system inside the ship. In the system, the flue gas emitted by the diesel engines passes through the EP to remove dust. The EP is energized by the high-voltage power supply with control panels, and the treated flue gas passes through the economizer where heat exchange between water and the gas occurs. Subsequently, the gas is introduced into the SO_x scrubber using seawater to remove SO_x. The structure of the SO_x scrubber is a shower type as explained in Subsection 3.3.2. The seawater is circulated using pumps.

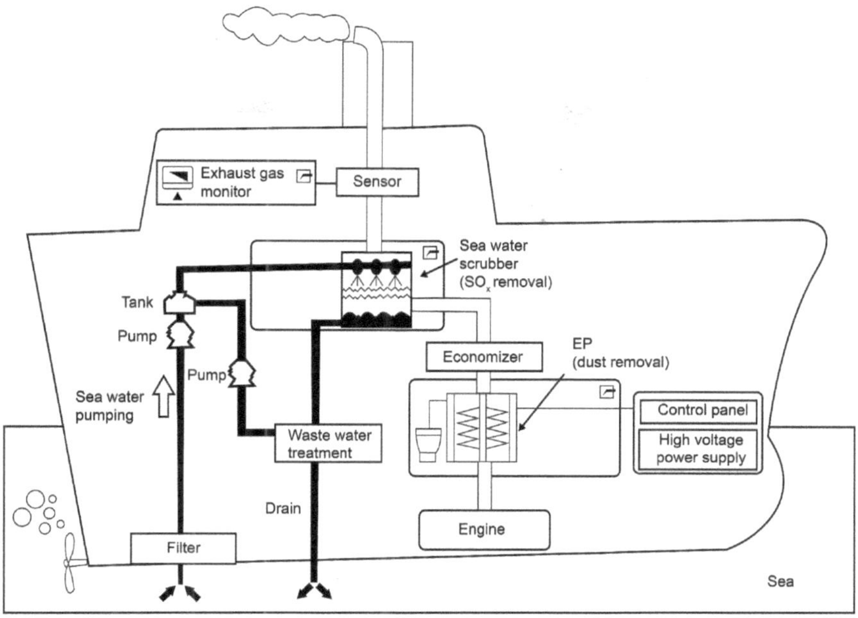

Figure 4.3.1 Concept of developed exhaust gas cleaning system developed inside a ship [10].

The wastewater is treated using the wastewater treatment apparatus. The seawater is supplied through filters using pumps. The exhaust gas is continuously monitored using gas sensors.

4.3.2 Example of marine diesel engine electrostatic precipitator

Inui et al. [10] developed a "hole-type EP" to avoid the reentrainment of collected PM. The principle of EP and reentrainment of low-resistivity dust for higher velocity conditions is explained in Chapter 3. The hole-type EP contains many holes in the collection plate electrode through which captured PM is passed. The captured PM accumulates inside the collection electrode to avoid the reentrainment of PM. When a large amount of PM is collected in the hole-type EP, the time-dependent collection efficiency or performance decreases as the effective area of the electrode decreases. Therefore, periodical automatic cleaning of the electrodes is necessary to remove the collected PM from the apparatus.

A dust compression machine is used to reduce the total volume of the PM by 70% compared with the raw collected PM. The final compressed PM is pellet-like and easily handled by workers. The main component of

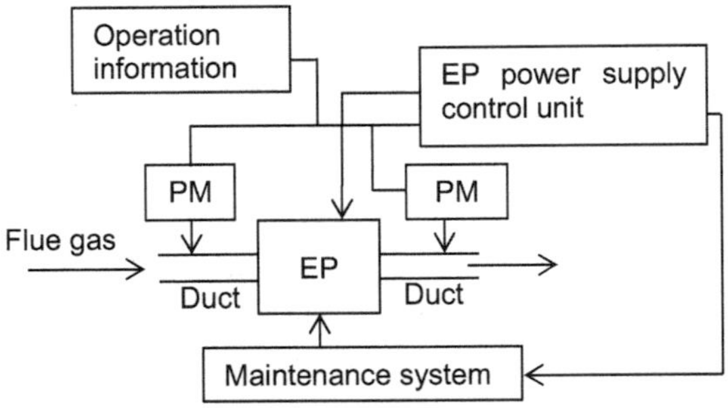

Figure 4.3.2 Example of a system for monitoring and controlling a marine diesel engine electrostatic precipitator (EP). *PM*, particulate matter. *(Made based on Ref. [10].)*

the recovered PM is dry soot or carbon; therefore, it may be used as fuels or can be recycled similar to fly ash from coal-fired boilers.

Periodical removal of the PM is required for the EP. The most simple and easy way to operate an EP is to maintain constant applied voltage or constant current and time-periodic maintenance or the removal of PM from the EP. However, the load of a marine engine often changes and the PM concentration varies. Therefore, an optimal operation to achieve the lowest operating cost of the EP and wastewater treatment could complicate the operation of the EP. In this regard, a novel monitoring and control system for marine diesel EPs was developed. Fig. 4.3.2 shows a schematic of the system.

In this system, the operation and performance of the EP are monitored by a pair of PM sensors shown as PM in the figure. These sensors, which are optical dust sensors, are used to continuously monitor the PM concentrations before and after the flue gas passes through the EP. When the PM concentrations reach certain threshold values, the control signals are directed to the EP and optimal operation can be achieved. Consequently, the collected dust is removed by the automatic maintenance mechanism.

Pilot tests were carried out with the hole-type EP and a laser-type dust monitor to assess the performance of the monitoring and controlling system for the EP. The specification of the tested diesel engines is as in Table 4.3.1. The experimental conditions are as follows. The internal temperature of the EP is 260–270°C and the internal velocity is approximately 9 m/s. The collection efficiency was continuously evaluated for 24 h. The EP was operated by applying constant voltage control of 7 kV. Automatic maintenance of a 1-h period was then carried out. Fig. 4.3.3 shows an example of the measurement result of the time-dependent collection efficiency. During operation, i.e., while the PM is collected inside the EP, a gradual

Table 4.3.1 Main component specification of the pilot plant test for electrostatic precipitator (EP).

	Specification	
Diesel engine	Power: 118 kW	Fuel: light oil
Compressor	Flow rate: 450 Nm^3/h	
	Temperature: 300°C	
	Fuel: light oil	
	W700 × D500 × H550	$V_{p-p} = 10$ kV
EP	Applied voltage: ~10 kV	
	Source: W240 × D100 × H160	FS
Dust monitor	Detector: W180 × D100 × H120	100 mg/Nm^3
	FS: 100 mg/Nm^3	

Made based on Ref. [10].

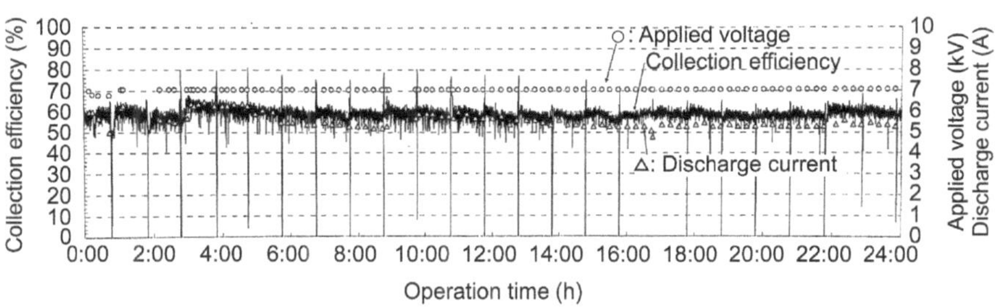

Figure 4.3.3 Measurement result of the collection efficiency of the monitoring and control system [10].

decrease in the collection efficiency is recognized, but the removal efficiency is restored by the automatic maintenance system. The experiment confirms that the removal efficiency is maintained at 55–60% during the 24-h period of operation.

4.4 Computational fluid dynamics for two-phase flows in electrostatic precipitators

4.4.1 Design of electrostatic precipitator by numerical simulation

Electrostatic precipitators (EPs) have a long history, and their invention can be traced back to the beginning of the 20th century. In recent years, almost all coal-fired power generation plants in Japan have been fully equipped with EPs, and they have certainly been a leading environmental tool in cleaning the country's skies. Their domestic rate of diffusion has been quite high, and they are now widely used in a variety of industries (thermal power stations, steel manufacturing plants, paper manufacturing plants, cement factories, chemical plants, glass melting furnace, etc.). However, owing to the growing awareness of the problem of nanoparticles, typified by the recent $PM_{2.5}$ problem, EPs that are low in pressure drop and high in nanoparticle collection efficiency have taken the spotlight in place of filters, are being applied to indoor air cleaners, and are being explored for use in automobiles.

We use three-dimensional two-phase flow in an EP as the subject, assume computational fluid dynamics (CFD) analysis using commercial simulation software, and explain a system of basic equations and solutions for electric fluid flow and particle motion [11]. After that, we introduce several simulation results based on the equations.

4.4.2 System of basic equations

Three-dimensional two-phase flow of an EP may be analyzed through simultaneous numerical analysis using three systems of equations [11—18]: Maxwell's equations for the electromagnetic field, a system of thermal fluid flow equations, and a particle momentum equation. Of these, electrical potential needs to be analyzed in a state of ions or other charged particles (electrical charge) distributed in space; thus, the equation for potential essentially becomes Poisson's equation. In addition, the momentum equation for the flow fields is the Navier—Stokes equation, which accounts for electrostatic forces. Particle and fluid flow has relative velocity and forms a two-phase flow. For the momentum equation for particles, two types of techniques exist. When we considered particle phases to be a continuum fluid, the Euler method is used, which is treated as a partial differential equation. When the force applied to a single particle is considered, the Lagrange method is used, which is treated as an ordinary differential equation. However, in practical analysis of EPs, the easily understood Lagrange method is often used, where the trajectories of individual test particles are analyzed. Normally, those that act on particles from surrounding fluid is considered; however, the impact of particles themselves and the reaction force exerted by particles on the fluid is assumed to be small in effect and is not considered. It should be noted that the physical variable of the electric field and flow field equations is the size of the field as a function of static coordinates (globally fixed coordinate system), while the physical variable of the particles is different and is a point acting as the physical quantity held by a single moving particle. The system of basic equations is shown below.

Basic equations for electric field

The Maxwell's equations (electromagnetic field equations) regularly used in EP analysis are equations for electric flux density $\boldsymbol{D}$ and current density $\boldsymbol{J}$.

$$\nabla\cdot\boldsymbol{D}=\rho_e,\quad \boldsymbol{D}=\varepsilon_r\varepsilon_0\boldsymbol{E} \tag{4.8), (4.9}$$

$$\frac{\partial\rho_e}{\partial t}+\nabla\cdot\boldsymbol{J}=0 \tag{4.10}$$

$$\boldsymbol{J}=\rho_e\boldsymbol{u}+\sigma\boldsymbol{E}=\rho_e(\boldsymbol{u}+b\boldsymbol{E})\approx\rho_e b\boldsymbol{E} \tag{4.11}$$

where ρ_e is the volume charge density, ε_r is the relative dielectric constant, ε_0 is the dielectric constant of vacuum, $\boldsymbol{u}$ is the fluid velocity, and b is the ion mobility. The value of b varies depending on the type of gas and whether ions are positive or negative, but can be regarded as a constant at constant temperature. When negative ions in the air are the target particles, $b=2.0\times10^{-4}\,\mathrm{m^2/(V\cdot s)}$ (20°C) in Ref. [13], $b=1.51\times10^{-4}\,\mathrm{m^2/(V\cdot s)}$ (0°C) in Ref. [18], and $b=2.5\times10^{-4}\,\mathrm{m^2/(V\cdot s)}$ (0°C) in Ref. [19]. In addition, ion velocity may be assumed to be large compared to u, and in that case an approximation of Eq. (4.11) holds. Potential (voltage) ϕ may be introduced by $\boldsymbol{E}=-\nabla\phi$, and Poisson's equation can be obtained from Eqs. (4.8) and (4.9).

$$\nabla\cdot(\varepsilon_r\varepsilon_0\nabla\varphi)=-\rho_e \tag{4.12}$$

In addition, when Eq. (4.11) is substituted into Eq. (4.10), the spatial discharge ρ_e is assumed to be stationary, and the following Eq. (4.13) is obtained. Furthermore, when the left side of Eq. (4.13) is expanded and Eq. (4.12) is considered, Eq. (4.14) is obtained.

$$\nabla\cdot(\rho_e b\nabla\varphi)=0\ \text{ or }\ \varepsilon_0\nabla\rho_e\cdot\nabla\varphi=\rho_e^2 \tag{4.13), (4.14}$$

By solving Eqs. (4.12) and (4.13) or Eq. (4.14), under the proper boundary conditions, the distribution of $\boldsymbol{E}$, ϕ, and ρ_e is determined.

Basic equations for thermal fluid

The basic equations of fluid (gas) flow within a machine are a continuity equation (mass conservation law) and a momentum equation or Navier–Stokes equation, and they are generally formulated as follows.

$$\frac{\partial\rho}{\partial t}+\nabla\cdot(\rho\boldsymbol{u})=0 \tag{4.15}$$

$$\frac{\partial(\rho \boldsymbol{u})}{\partial t} + \nabla\cdot(\rho \boldsymbol{u}\boldsymbol{u}) = -\nabla p + \nabla\cdot(\mu\nabla \boldsymbol{u}) + \frac{1}{3}\mu\nabla(\nabla\cdot \boldsymbol{u}) + \rho_e \boldsymbol{E} \tag{4.16}$$

where ρ is the fluid density, t is the time, p is the pressure, and μ is the coefficient of fluid viscosity. Electrostatic force as the volume force acting on the fluid is considered, which is a source of secondary flow due to ion wind. The energy equation for determining fluid temperature T (equation of state and law of the conservation of energy) is as follows.

$$p = \rho RT \tag{4.17}$$

$$C_p\left[\frac{\partial(\rho T)}{\partial t} + \nabla\cdot(\rho \boldsymbol{u} T)\right] = \nabla\cdot(\lambda\nabla T) + \psi_D + S_c \tag{4.18}$$

where R is the gas constant, T is the absolute temperature, C_p is the specific heat at constant pressure, λ is the thermal conductivity, ψ_D is the dissipative energy loss, and S_c is the source term including the energy for the chemical reaction. The distribution of $\boldsymbol{E}$, ϕ, and ρ_e are determined from Eqs. (4.12) to (4.14), and by solving Eqs. (4.15)–(4.18) from this distribution, flow field $\boldsymbol{u}$ and temperature T are determined.

Particle electric charge equations

The key factor for an accurate simulation of EP collection efficiency is the estimates of the amount of particle electric charge. If particles larger than 1 μm in diameter are the target particles, then the so-called the field charging is dominant. The mechanism of this is ions moving along an electric field colliding with particles, adhering to them, and then electrically charging them. The amount of electric charge gradually increases, achieving saturation charge q_{max}. The field charging process may be expressed by the relaxation equation below [16], as already explained in Subsection 3.4.3.

$$\frac{dq}{dt} = \frac{1}{\tau q_{max}}(q_{max} - q)^2 \tag{4.19}$$

$$q_{max} = \frac{3\pi\varepsilon_0\varepsilon_r|\boldsymbol{E}|d_p^2}{\varepsilon_r + 2} \tag{4.20}$$

where q is the amount of particle electric charge, and $\tau = 4\varepsilon_0/(\rho_e\, b)$ is the relaxation time until the particle electric charge has reached 50% saturation. Because this is extremely small at $\tau = 10^{-3}$ s, q may be immediately reaching saturation on entering an EP.

If nanoparticles of approximately $d_p < 100$ nm are the target particles, then diffusion charging is dominant. The change over time of the amount of particle charge is expressed by the following equations [19] as already explained in Subsection 3.4.3:

$$q(t) = q^* \ln\left(1 + \frac{t}{\tau^*}\right) \tag{4.21}$$

$$q^* = \frac{2\pi\varepsilon_0 d_p kT}{e} \tag{4.22}$$

where k is Boltzmann's constant and e is the elementary charge. Although, theoretically, the amount of saturation charge q_{max} is not determined by Eq. (4.21), $q = 6.2q^*$ at $t = 500\tau^*$ is used as a value for q_{max} [19].

Particle momentum equation

A momentum equation for simulating the motion of one test particle may be written as follows, according to the Lagrange method.

$$m_p \frac{d\boldsymbol{u}_p}{dt} = \frac{C_D}{C_C}\rho(\boldsymbol{u} - \boldsymbol{u}_p)|\boldsymbol{u} - \boldsymbol{u}_p|\frac{A_p}{2} + m_p\boldsymbol{g} + q\boldsymbol{E} + \boldsymbol{F}_{Ba} + \boldsymbol{F}_{Saff} + \boldsymbol{F}_p + \boldsymbol{F}_{Br} \tag{4.23}$$

where m_p is the mass of a single particle, $\boldsymbol{u}_p$ is the particle velocity vector, C_D is the drag coefficient, C_C is Cunningham's correction factor, ρ is the surrounding fluid density, A_p (for spherical particle, $A_p = \pi d_p^2/4$) is the particle projection area, and $\boldsymbol{g}$ is the gravity vector. On the right side of Eq. (4.23) are given the drag force, gravity force, electrostatic force, and the following unsteady drag forces: $\boldsymbol{F}_{Ba}$, the Basset force, $\boldsymbol{F}_{Saff}$, the Saffman lifting force, $\boldsymbol{F}_p$, the force due to pressure gradient including buoyancy, and $\boldsymbol{F}_{Br}$, the force due to Brownian diffusion. Furthermore, the inertial force due to virtual mass is small in solid-gas two-phase flows; thus, it generally may be ignored.

On the right side of Eq. (4.23), the influence of drag force, gravity, and electrostatic force are generally significant. The drag force is influenced not only by particle shape and direction but also by particle Reynolds number, turbulence level, and other factors. For nanoparticles having sizes near the mean free path of gas fluid molecules, particle slip occurs. As a result, the drag force is needed to be corrected. As shown in Eq. (4.23), corrections may be applied by dividing C_D by C_C (the value of which is larger than one). Furthermore, $\boldsymbol{F}_{Ba}$, $\boldsymbol{F}_{Saff}$, $\boldsymbol{F}_p$, and $\boldsymbol{F}_{Br}$ ought to be properly considered in the precise calculation.

Boundary conditions

The boundary conditions for solving the aforementioned fundamental equations must consider two types of boundary conditions: (1) thermofluid dynamic boundary conditions, such as a zero velocity on wall surfaces, constant velocity gradient, constant temperature, constant heat flux, and free in- and outflow conditions and (2) electromagnetic field boundary conditions, which are imposed when the potential ϕ or another value for a physical quantity is conferred on a boundary for electrofluid dynamics or when there is a boundary surface consisting of two mediums with different conductivity values within a system, wherein the normal component of current density $\boldsymbol{J}$ is continuous, and the tangential component of electric field $\boldsymbol{E}$.

Boundary conditions should be given in accordance with actual systems and are difficult to describe in a unified manner. In the following, an outline of them for individual analytical cases is provided.

Analytical procedure

The equations governing electrohydrodynamics described above are numerically solved as a coupling problem of the differential equations, and calculations are made to obtain electric fields, electrical charge fields, velocity fields, temperature fields, and particle trajectories. Electric fields, electrical charge fields, velocity fields, and temperature fields comprise a system of simultaneous partial differential equations; however, electric fields and electrical charge fields may be determined independently, and they pose relatively easy problems for coupling computational hydrodynamics. In addition, in EP analysis, it is often assumed that flow fields are stationary, temperature is constant (with no consideration of changes in gas temperature), and Eqs. (4.17) and (4.18) are not analyzed. In addition, if flow velocity or Reynolds number becomes larger, the main flow becomes turbulent, and computational grid near the flow path is not extremely fine, then an accurate solution will not be achievable with direct simulation. Therefore, the velocity distribution is interpolated near a wall with a wall function or something similar, and then turbulence models such as $k-\varepsilon$ model, etc., are used to consider turbulent viscosity. In addition, when the flow reaches high speeds close to the speed of sound, the influence of incompressible is considered; therefore, normally, analysis is done with incompressible models of constant fluid density.

For two-phase flow calculations, for example, CFD-ACE+ [11], FLUENT [15,16], and STAR-CD [17,18], are well-known commercial solvers. If these are used, the simulations proceed with various calculation conditions stipulated in the software for solvers.

4.4.3 Analysis results

Thereafter, using the above system of basic equations, analysis results are shown. For the sake of calculation cost and other factors, approximations are performed in each analysis, where all terms in the aforementioned fundamental equations are always considered; however, terms that are of negligible influence are ignored. Readers interested in more detailed explanations should be prompted by this text to consult the relevant papers.

Three-dimensional analysis of ion wind secondary flow within tuft/point corona electrostatic precipitator

In the years since the implementation of EP, wire (piano wire) discharge electrodes and flat plate collecting electrodes have been adopted; however, dust adheres to discharge electrodes, often leading to decreased collection performance problems because of insufficient current or discharge electrode disconnection accidents. In recent years, EPs have increased owing to an increase in the amount of processing gas; furthermore, there is an increased demand for improved performance and machine reliability and a general adoption of point (projected)-type discharge electrodes (tuft/point corona discharge wire) in which current easily flows. In particular, with submicron high-density particle collection in the exhaust gas of high sulfur-oil-fueled boilers, long projected-type discharge electrodes are essential for dealing with decreased performance due to increased dust collection spatial discharge effects.

The results of analyzing EP flow within such EP [13,14] are introduced. Flow is the superimposition of the main flow and secondary flow, which is called ion wind in static electric fields. As a first step, Eqs. (4.12) and (4.13) are analyzed by the successive relaxation method using the finite differential method and obtained the distribution of $\boldsymbol{E}$, ϕ, and ρ_e. Specifically, the procedure below is followed. Cooperman's solution is used, which is a solution of Laplace equation [13,14] as an initial condition. The obtained potential ϕ is differentiated and the initial values of the electric field intensity E_x, E_y, and E_z in the main flow direction x, a direction perpendicular to the collecting plate electrode y, and a direction parallel to the collecting plate electrode z are determined. Using these initial values and current density set values, values for spatial discharge density near tuft/point corona are obtained by Ohm's law equation Eq. (4.11). Furthermore, using the obtained values for spatial discharge density, potential (voltage) is calculated from Poisson's equation Eq. (4.12). When this is done, calculations are repeated until the difference between the newest value of voltage and voltage value one before at each lattice point is within 0.01%. Finally, if current density values are compared with set values, and the difference does not match within 0.01%, the calculation is repeated. Assuming a constant flow field temperature, only the continuity Eq. (4.15) and momentum Eq. (4.16) are analyzed. Furthermore, a $k-\varepsilon$ model, as a turbulence model, is used.

Fig. 4.4.1 shows the results of analyzing a flow field excited by ion wind without main flow. The line in the figure represents the flow line, and the differences in color represent differences in positions in the y-axis direction in the dust collection space. In addition, in Fig. 4.4.1A and B, flow in an $x-z$ plane including a point corona is symmetrical; thus, only the top flow field is displayed. Donut-shaped ring flow originating from a point discharge electrode, blowing out radially in the collecting electrode direction and gas flow direction and also returning in the discharge electrode direction, occurs in the dust collection space between point discharge electrodes. Fig. 4.4.2 shows a schematic depiction of the flow regime for the ring flow.

Fig. 4.4.3 shows the results of analyzing a flow field with a main flow present, as well as nonlinear interaction with ion wind. As the main flow

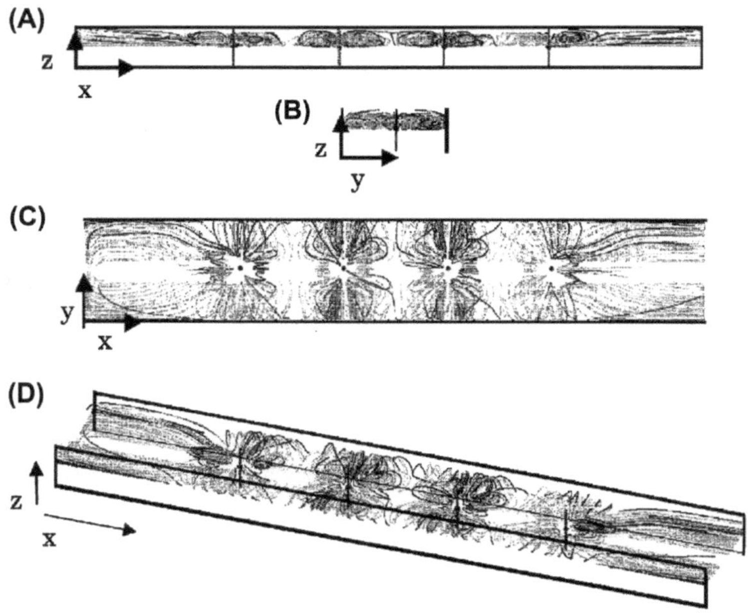

Figure 4.4.1 Flow in the collecting region without main [13]. (A) $x-z$ cross-sectional view, (B) $y-z$ cross-sectional view, (C) $x-y$ cross-sectional view, and (D) 30-degree inclined view. $U = 0.0$ m/s and $N_{ehd} = \infty$.

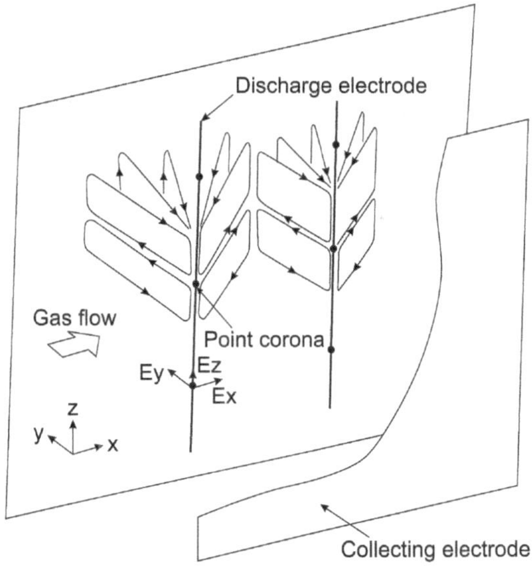

Figure 4.4.2 Characteristics of three-dimensional secondary flow near the point discharge electrodes.

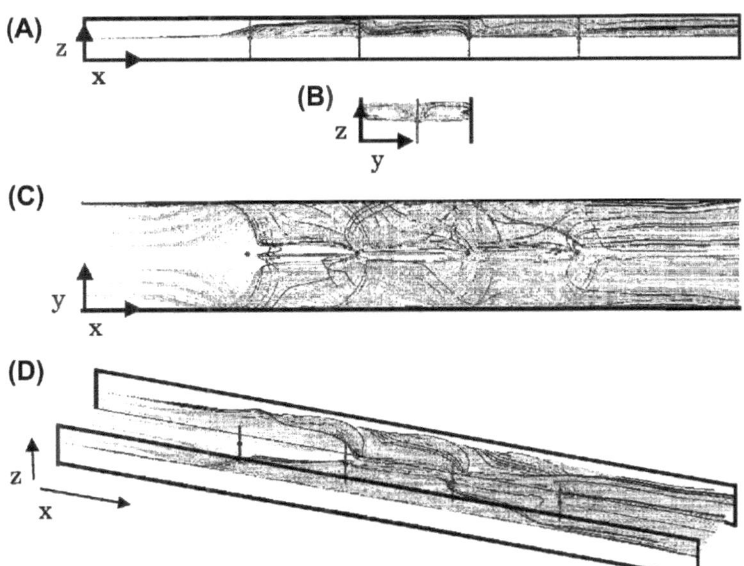

Figure 4.4.3 Turbulent flow in the collecting region [13]. (A) x–z cross-sectional view, (B) y–z cross-sectional view, (C) x–y cross-sectional view, and (D) 30-degree inclined view. $U = 0.5$ m/s and $N_{ehd} = 0.477$.

velocity becomes larger, ring flow is suppressed as shown in Fig. 4.4.2 and moves in the downstream direction; overall flow starts to rotate; and three-dimensional spiral flow is formed. Then, when main flow velocity is approximately 0.5 m/s, three-dimensional spiral flow is suppressed as shown in Fig. 4.4.3, becomes diffused, and turns into mixed flow in the outlet. It is believed that such flow may be formed in an actual EP dust collection space.

The following electrohydrodynamics (EHD) number N_{ehd} is assumed to be a dimensionless number, representing the influence of secondary flow (ion wind) flow velocity u_e on the mainstream of gas U.

$$N_{ehd} = u_e/U, \quad u_e = \sqrt{J_p d/(b\rho)} \tag{4.24}$$

where U is the main flow typical velocity, J_p is the collecting plate current density immediately below point corona, d is the electrode interval between discharge electrode and collecting electrode, ρ is the fluid density, and b is the aforementioned ion mobility. In the results of Fig. 4.4.3, it is assumed that $u_e = 0.238$ m/s ($J_p = 0.12$ mA/m^2, $d = 0.114$ m, $b = 2.0 \times 10^{-4}$ m^2/(V·s), and $\rho = 1.205$ kg/m^3 (20°C)) and $U = 0.5$ m/s, and therefore $N_{ehd} = 0.477$. However, in actual dry-type EP for coal-fired boilers, $J_p = 0.2-0.35$ mA/m^2, $d = 0.15$ m, and $U = 1.0$ m/s are common values, thus $N_{ehd} = 0.35-0.47$, and although current density and velocity values are different, the results in Fig. 4.4.3 may be said to be ones in which relative flow is calculated for an actual machine. Furthermore, according to Adachi et al. [20], when the EHD number exceeds 2, ion wind becomes larger, and flow interaction between ring flow and main flow becomes large. However, under general operating conditions for this type of EP, EHD number is at most approximately equal to 1, and it is believed that ion wind does not have much influence.

Effect of gas temperature on collection efficiency in electrostatic precipitator

The temperature of exhaust gas targeted by EPs sometimes reaches 200°C or higher, and few studies have been reported systematically examining the influence of such high exhaust gas temperature on particle behavior, secondary flow situations, and decreased collection efficiency. Here, the results of examining this through numerical simulation [15] are reviewed.

As exhaust gas reaches high temperatures, ion wind strength increases, and the secondary flow becomes pronounced. This is mainly due to the mechanism below. Ion electric charge density increases overall as temperature rises, and at high temperatures, the viscosity coefficient of gas becomes larger; hence, there is a stronger tendency for the force experienced by the particles in the flow and electric fields (drag force and electrostatic force) to become larger and move along the flow fields. According to the momentum conservation law of gas flow, the influence of secondary flow may be ignored with low mass flow rates or low inlet flow velocity. However, the larger the influence of secondary flow on the collection, the smaller the size of the targeted particles, at the point at which the influence on collection efficiency becomes pronounced. Particle trajectories, when particles are collected between electrodes, are precisely simulated in calculation results for whether or not there is secondary flow, and results are reported in which secondary flow has a greater influence when gas temperatures are high, particle densities are large, and particle sizes are small. In addition, there are two types of charge principles for particles: one for electric charge and one for diffusion charging. The former weakens as temperature rises, whereas the latter becomes stronger. Accordingly, it is considered that collection efficiency increases for particles that are of small size and the influence of diffusion charging becomes large.

Three-dimensional flow simulation in actual shape electrostatic precipitator

In actual EP, it has a complicated shape such as a rectifying plate for rectifying the flow at the inlet, hoppers for collecting the collected fine particles, and a plurality of collecting electrodes. In this text, the calculation of three-dimensional flow and fine particle collection of complex practical shaped wire-electrode type EP [16] are introduced. Fig. 4.4.4 shows a diagram of the three-dimensional EP having the targeted practical shape. Effective length, effective width, entrance height are 0.45 m, 0.3 m, and 0.45 m, respectively; three rectifying plates (air distribution plate) exist in it, and the distances from the inlet are 0.3 m, 0.5 m, and 0.65 m. The length of the bell-type flow channel at the entrance is 0.65 m. The collecting box has 3.2 m in length, 1 m in width, and 1.54 m in height and is equipped with 3 groups of 12 dust-collecting plates and 15 negative discharging lines. The walls on both side surfaces of the collection box are also used as a dust collection plate. The distance between the discharge wire and the electrode plate is 0.1 m, the wire interval is 0.2 m, and the length and height of the dust collecting plate are 0.8 and 1.5 m. The length of the outlet is 0.3 m, the width 0.3 m, and the height 0.3 m. The length of the outlet bell-type flow path is 0.45 m, the height of the three dust collection hoppers is 1 m, and the calculated mesh of EP is 888,245 corresponding to 798,950 nodes.

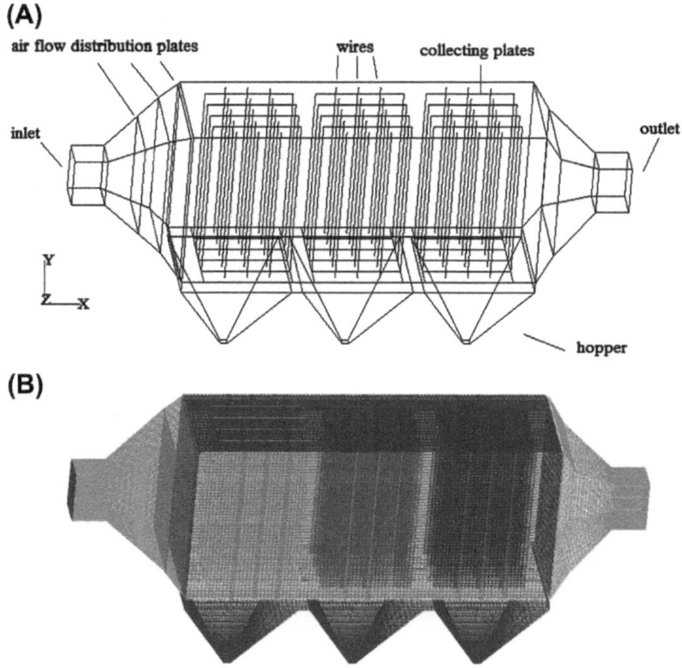

Figure 4.4.4 Three-dimensional calculation result [16]. (A) Duct shape and (B) reproductive mesh.

Fig. 4.4.5 shows the trajectories of 100 test particles (particle diameter = 5 μm and particle concentration = 0.001 kg/m^3) for different applied voltages. In the case of 30 kV, the particle spreads throughout the entire area, but in the case of 70 kV, the particles are almost completely collected (collection efficiency ~ 98%) in the upper region and the central region excluding a part of the lower region of the flow stream near the vicinity of the hopper. In this chapter, it is shown that for a particle of 1 μm under the same condition, the collection efficiency decreases down to approximately 60% because of the difficulty in charging. Furthermore, if the flow rate is increased up to 20 m/s, the influence of turbulence increases and the collection efficiency decreases. As an example, it is shown that the collection efficiency decreases to approximately 30% at 40 kV. On the other hand, although the particulate concentration does not affect the collection efficiency, the repelling forces to the fluid become large and could affect the flow patterns.

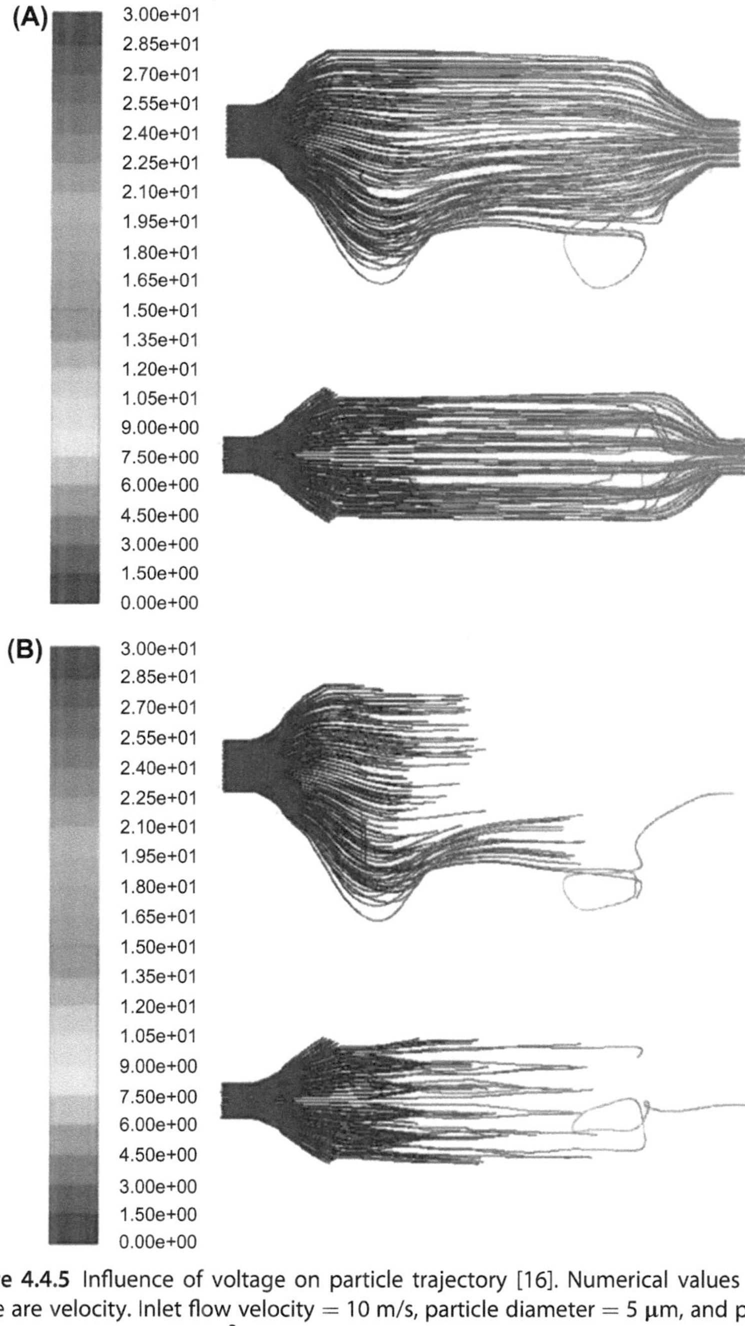

Figure 4.4.5 Influence of voltage on particle trajectory [16]. Numerical values in the figure are velocity. Inlet flow velocity = 10 m/s, particle diameter = 5 μm, and particle concentration = 0.001 kg/m^3. (A) Results are shown in case of 30 kV and (B) in case of 70 kV.

Simulation of ionic wind, secondary flow, and fine particle collection in coaxial cylinder shape electrostatic precipitator

Hole-type EP was devised for the purpose of preventing rescattering of particulates with low electrical resistivity (<10^4 Ω cm) such as diesel engine exhaust particulates from the collecting electrode plate. Because particles enter the pocket on the outer circumference of the grounded electrode from the hole, it makes it possible to prevent scattering again. Numerical simulation results for hole-type negative-polarity needle-type EP are introduced [17,18].

Fig. 4.4.6 shows a calculation example of three-dimensional ion wind. This EP has a coaxial cylinder shape, needle-type electrodes at the center is used for negative polarity discharge electrode, and a hole-type electrode (diameter = 57.5 mm) with holes (diameter = 2.5 mm) is used as a grounded electrode. The total length is 500 mm. A negative direct current high voltage of −7.8 kV is applied between an internal grounded electrode with holes and centered high-voltage electrode. In this calculation example, it is assumed that the main flow velocity in the EP is zero. In the calculation result, it can be observed that ionic wind or corona wind is generated from

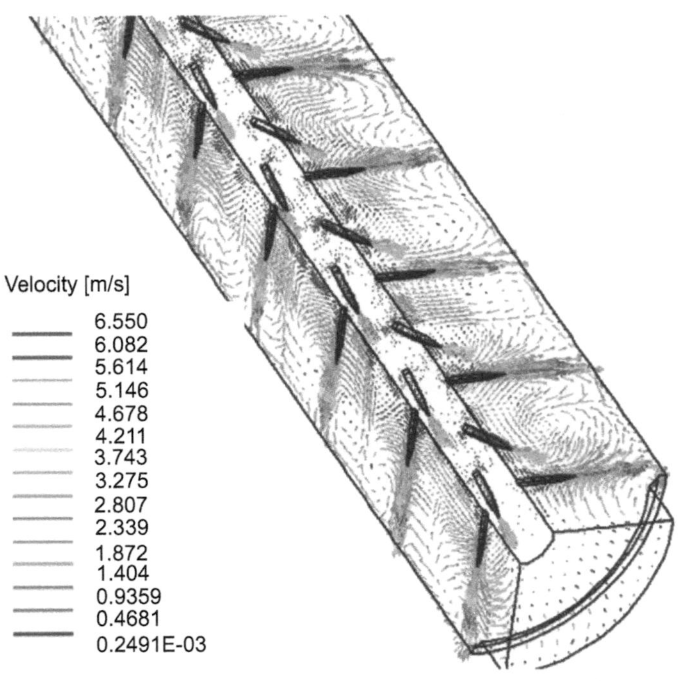

Figure 4.4.6 Distribution of three-dimensional ion wind [18].

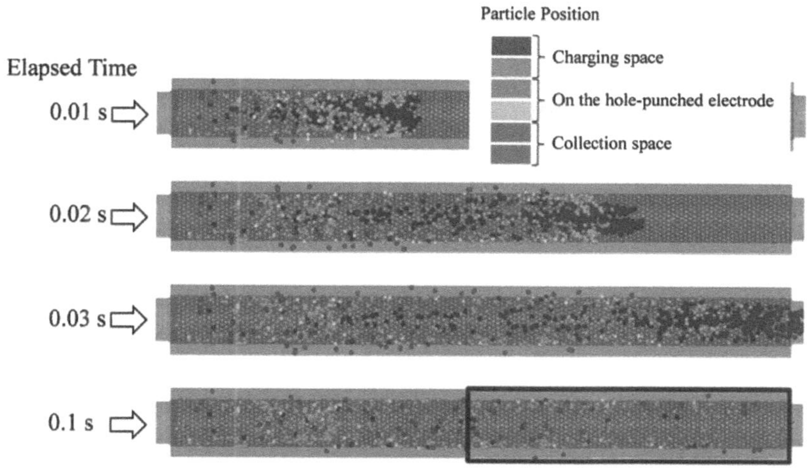

Figure 4.4.7 Particle movement in the presence of mainstream and secondary flow [18].

all the needle electrodes. A circulating flow toward the direction of the ground electrode with the holes from the needle electrodes is induced, and the flow is returning to the center. It is a typical ionic wind characteristic, and the flow velocity is a maximum of 6.5 m/s and a minimum of 0.25 m/s.

Fig. 4.4.7 shows the time-dependent movement of the particles (diameter = 10 μm) when there are main flow and ion wind. When considering the ion wind in the calculation, especially in the downstream part for the result at $t = 0.1$ s surrounded by the square in the figure, the secondary flow in the radial direction becomes significant, and therefore, the trapping rate improves. In the calculation, the total collection efficiency is 45.7% when considering the secondary flow, while it decreases down to 36.9% when not considering the secondary flow.

4.4.4 Application of electronic precipitators to marine diesel engines

The fundamental equations for analyzing fluid flow and particle motion, analytical methods, and the results of analyzing fundamental equations regarding three-dimensional two-phase fluid simulation of an EP are explained. EPs will become the main PM removal technology in the future. EPs are being explored for future use because of their superior nanoparticle collection with little ventilation pressure drop. This is not only for large-scale EPs in thermal power stations but also, for example, for PM removal equipment in diesel automobiles currently using ceramic filters.

Studies on the application of EPs to marine diesel engines are also progressing. Nonetheless, the problem of reentrainment of low resistivity fine particles cannot be said to be completely solved. However, countermeasures such as hole-type EP [17,18] have reached their practical application stage. Here, hole-type EP refers to those devised for the purpose of preventing reentrainment from collecting electrode plates for particles that are low in electrical resistivity ($< 10^4\ \Omega$ cm), such as diesel engine exhaust particles, which enable prevention of reentrainment by having collected particles brought from holes into pockets on the outer circumference of ground electrodes. This technology is very promising and will contribute the electrostatic precipitation for marine diesel engine emission control.

4.5 Laboratory-scale plasma particulate matter decomposition

4.5.1 Diesel particulate filter regeneration

Diesel particulate filter (DPF) regeneration corresponds to the PM removal deposited inside it. The technology to be done is important to realize a clean diesel engine. The principle is explained in Chapter 3. DPF regeneration using plasma-induced ozone injection is explained here based on Ref. [39] in Chapter 3. Fig. 4.5.1 shows a schematic of the experimental apparatus for DPF regeneration using ozone injection. The engine that is used for the investigation is a diesel engine generator (maximum load = 2 kW,

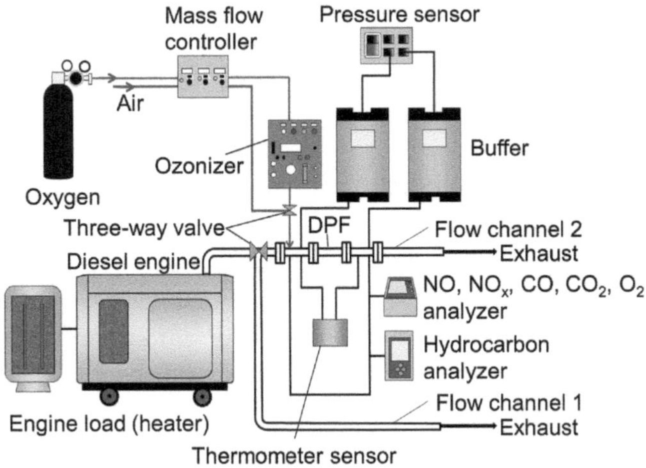

Figure 4.5.1 Schematic of experimental apparatus for diesel particulate filter regeneration.

displacement = 219 mL, single cylinder type, and rotation rate = 3600 rpm; YDG200VS-6E, YANMAR Co., Ltd.) that uses gas oil or light diesel oil (sulfur concentration = 7 ppm) as fuel. During the experiment, the flow rate of the exhaust gas is maintained at 300 NL/min, where N means the standard state (0°C and 101.325 kPa), and a load in the form of a 1-kW electric heater is connected to the power output terminal. Before the experiment, the engine is warmed up for 1 h with the exhaust gas emitted through flow channel 1. Immediately after warming up, the experiment is performed by switching the exhaust flow via a valve from flow channel 1 to flow channel 2, in which a SiC DPF (diameter = 50 mm, length = 80 mm, cell density = 300 cpsi ≈ 46.5 cells/cm^2, wall thickness = 12 mil ≈ 0.3 mm, and material: SiC, TYK Corporation) is placed. The PM is initially captured in the DPF for some time. Then, without stopping the engine, O_3 generated by a plasma-induced ozonizer (surface discharge type, OZS−HC−70/W, Masuda Research Inc.) is injected at the inlet of the DPF for the regeneration. DPF regeneration is evaluated by measuring the pressure differences. Table 4.5.1 shows the experimental conditions of the flow rate of the supplied O_2 for the ozonizer, the mass flow rates, and concentrations of O_3. The ozone is produced by supplying O_2 of 99.6% purity from a gas cylinder to the ozonizer with an input power of 370 W. The ozonizer produces and emits O_3 as a mixture of O_3 and O_2. An ozone injection tube is installed at the center of the cross section of the flow channel and 35 mm in front of the DPF inlet against the exhaust gas flow. O_3 is injected into the DPF apparatus in a direction opposite to the exhaust gas flow. The PM collection is set to 1−2 h; the DPF regeneration is set to 2−5.5 h; and the O_3 injection rate is 25.8−34.8 g(O_3)/h (the O_3 concentration after injection is 658−873 ppm). During PM collection, the injected flow rate of air is maintained equal to that of O_3 during DPF regeneration.

The engine load is set to 1.0 kW (50% of the maximum output power) for all the experimental trials. The pressures at the inlet and outlet of the DPF are measured using pressure sensors (amplifier AP-C40, sensor head AP-44, KEYENCE Co.). The pressure difference between the inlet and

Table 4.5.1 Experimental condition.

Volumetric flow rate of supplied O_2	L/min	5	8	10
Mass flow rate of O_3	g/h	25.8	32.9	34.8
Concentration of O_3 at outlet of ozonizer	%	4.10	3.25	2.74
Concentration of O_3 after injection	ppm	658	831	873

outlet channels is determined from the measured pressures. The pressure sensor has two resolutions—0.02 and 0.1 kPa—for different pressure ranges. A resolution of 0.02 kPa is used at both the inlet and the outlet when $\Delta p \leq 14$ kPa, whereas 0.1 kPa is used at the inlet and 0.02 kPa is used at the outlet when $\Delta p > 14$ kPa. In addition, the exhaust gas temperatures upstream and downstream of the DPF are measured using thermocouples (type K) connected to a recorder (NR-1000, KEYENCE Co.). Gas components such as NO, NO_x (=NO + NO_2), CO, CO_2, O_2, and hydrocarbons (HCs) upstream and downstream of the DPF are also analyzed using gas analyzers (PG-240, Horiba, Ltd., Japan and VMS-100F, Shimadzu Co., Japan). The pressure and temperature measurements and the gas sampling are taken at specified points, namely, 95 mm from the DPF inlet pipe and 145 mm from the outlet pipe.

When the pressure difference returns to the initial value before the PM collection, the DPF is considered to have been completely regenerated, in other words, complete DPF regeneration has been achieved. It should be noted that the DPF and flow channel are covered with thermal insulation material to minimize the effect of variations at ambient temperature. The mass of the collected and oxidized PM is determined by measuring the mass of the DPF using an electromagnetic force balance (PFII-3000, Shinko Denshi Co., Ltd., measurement accuracy = 10 mg) after it is gradually cooled to between 18 and 19°C.

4.5.2 Results of regeneration

Fig. 4.5.2 shows the results of the seventh and eighth trials of the DPF regeneration experiment out of a total of eight trials. The pressure differences Δp are indicated by the open circles. The upstream and downstream gas temperatures T_{in} and T_{out} are indicated by the open and solid triangles, respectively. The durations of the PM collection and regeneration of the seventh trial are 1 and 5 h, respectively, while they are both 2 h for the eighth trial. In the eighth trail, the quantity of PM captured in the DPF is 0.81 g (the PM deposition rate is 0.41 g(PM)/h). In the elapsed time of 5 h of the seventh trial, the pressure difference Δp shifts from increase to decrease when the O_3 injection rate is increased from 25.8 to 32.9 g(O_3)/h. In other words, increasing the O_3 injection rate enhances the PM oxidation and hence the DPF regeneration. In both the seventh and eighth trials, the pressure difference Δp decreases to 2.06 kPa. This is the target (initial) value

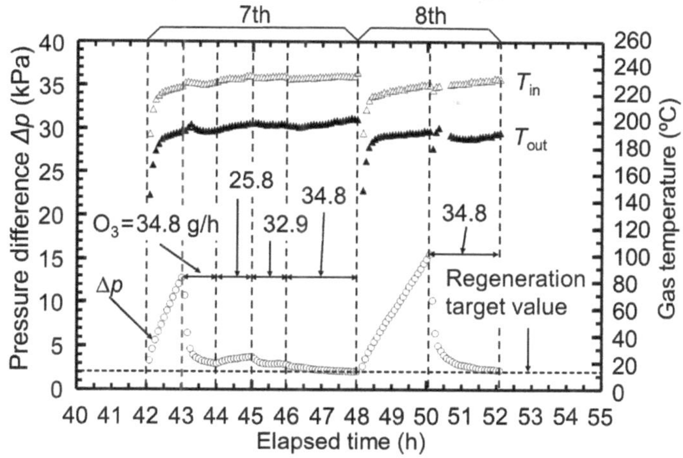

Figure 4.5.2 Pressure difference and gas temperature versus elapsed time in diesel particulate filter regeneration for the seventh and eighth trials for O_3 injection rate = 25.8–34.8 g/h.

when all PM is removed at the end of each regeneration cycle. In other words, complete DPF regeneration is achieved in each of these trials. Comparing the values of Δp during the first hour of O_3 injection, a decrease from 12.76 to 2.94 kPa (9.82 kPa decrease) is observed in the seventh trial, whereas a decrease from 15.5 to 2.76 kPa (12.74 kPa decrease) is observed in the eighth trial, for an O_3 injection rate of 34.8 g/h. The rate of PM removal is known to increase with an increase in the deposited mass.

The typical gas components of the PM collected downstream of the DPF are $NO = 224$ ppm, $NO_2 = 37$ ppm, $CO = 424$ ppm, $CO_2 = 6.2\%$, $O_2 = 13.1\%$, and the total gaseous HC concentration = 162 ppmC (i.e., the concentration based on the number of carbon atoms). In the case of DPF regeneration by O_3 injection, the average gas components are $NO = 4$ ppm, $NO_2 = 27$ ppm, $CO = 463$ ppm, $CO_2 = 6.2\%$, $O_2 = 14.6\%$, and the total gaseous HC concentration = 98 ppmC. The observed increase in the CO content after the O_3 injection confirms that the soot is oxidized. Regarding the observed decrease in the total HC concentration, not only the particulate HC compounds but also the gaseous HCs are oxidized by the ozone injection. Although the NO and NO_2 contents also decrease as a result, the detailed mechanism for this decrease is not yet understood.

Fig. 4.5.3A and B show photographic images of the DPF inlet in the eighth trial, before and after complete regeneration, respectively.

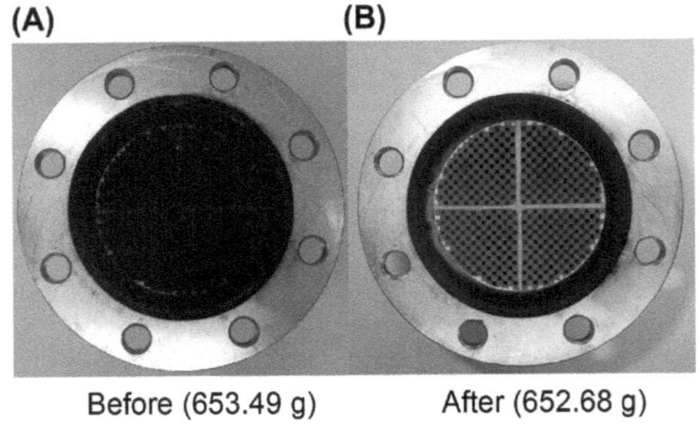

Figure 4.5.3 Photographic image of diesel particulate filter. (A) Before and (B) after regeneration for 2 h for O_3 injection rate = 34.8 g/h.

Although the PM accumulates in the DPF before the regeneration, as can be observed in Fig. 4.5.3A, the entire DPF surface appears clean in Fig. 4.5.3B, indicating that the accumulated PM has been oxidized by the injected O_3. The mass of the DPF also changes from 653.49 g (after 2 h of PM deposition and before regeneration) to 652.68 g (after complete regeneration), which also indicates oxidation of the PM. Furthermore, it should be noted that the mass of the DPF after the seventh trial is 652.68 g, which is equivalent to the DPF mass after complete regeneration in the eighth trial. This further confirms the achievement of complete regeneration in both the seventh and the eighth trials.

4.5.3 Analysis of regeneration

The regeneration characteristics are examined by analyzing the PM incineration by O_3 injection. Although the PM exists in the solid state, it is treated as a gas in the analysis. In addition, it is assumed that the PM oxidation is mainly caused by the existence of O_3. The reaction between the PM and the O_3 is experimentally known as of the second order and thus the reaction rate is given by

$$\textit{reaction rate} = k[\mathrm{PM}][\mathrm{O_3}]^2 \tag{4.25}$$

Denoting the concentrations of the PM and O_3, namely, [PM] and [O_3], by y and y_1, respectively, when the PM collection by the DPF is considered, Eq. (4.25) becomes

$$\frac{dy}{dt} = B - kyy_1^2 \tag{4.26}$$

where B (> 0) is a constant parameter of the PM collection effect. Moreover, assuming $[O_3] = y_1 =$ constant and employing $k\ [O_3]^2 = k'$, the above equation can be rewritten as follows:

$$\frac{dy}{dt} = B - k'y \tag{4.27}$$

The PM concentration y is also given by

$$y = \frac{N_A m}{MV} = \frac{m}{F}, \quad F = \frac{MV}{N_A} \tag{4.28}$$

where N_A is the Avogadro constant (6.022×10^{23} mol^{-1}), m is the time-dependent mass of the PM deposited in the DPF, M is the mean molecular mass of the PM, and V is the volume of the DPF ($=1.57 \times 10^{-7}$ m^3). It is estimated that $M = 8$ g/mol because the PM contains HC compounds, and their mean molecular mass would, therefore, be lower than that of carbon. As a result, $F = 2.09 \times 10^{-30}$ g m^3. Moreover, the relationship between Δp and m is considered to be linear:

$$m = A\Delta p \tag{4.29}$$

where A is a constant parameter determined to be 6.03×10^{-2} g/kPa from the measurements taken from the collected PM of mass 0.81 g for $\Delta p = 13.44$ kPa during the eighth trial.

During the PM collection process with the NTP off, k' is zero. Therefore, we have

$$\frac{dy}{dt} = B \tag{4.30}$$

$$y = B(t - t_0) + y_0 \quad (t_0 \le t \le t_1) \tag{4.31}$$

$$m = FB(t - t_0) + m_0 \quad (t_0 \le t \le t_1) \tag{4.32}$$

where y_0 is y at the time t_0, and t_0 and t_1 denote the beginning and the end, respectively, of the PM collection.

During the PM incineration process with the NTP on, k' is not zero. Therefore, we have the following solution to Eq. (4.27):

$$y = \frac{B}{k'} + Ce^{-k'(t-t_1)} \quad (t_1 \le t \le t_2) \tag{4.33}$$

where t_2 is the elapsed time at the end of the PM incineration, and B and $C > 0$ are constant parameters. From Eqs. (4.28) and (4.33), during DPF regeneration, m is given by

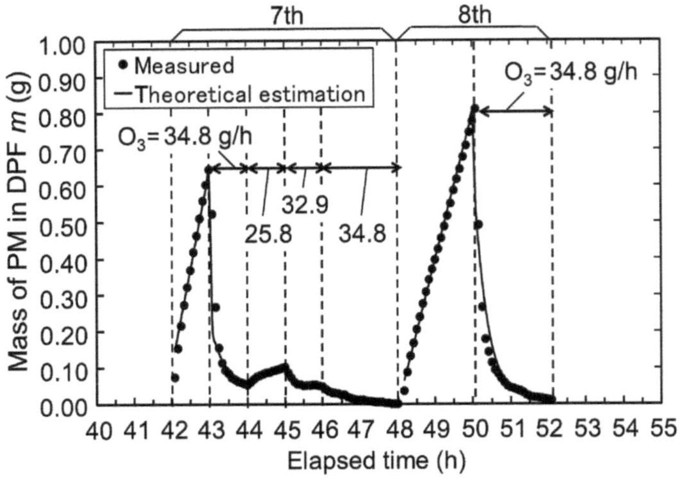

Figure 4.5.4 Time-dependent mass of PM in diesel particulate filter (DPF) from experimental results and theoretical estimation (O_3 injection rate = 34.8 g/h) for seventh and eighth trials of DPF regeneration. *PM*, particulate matter.

$$m = F\left[\frac{B}{k'} + Ce^{-k'(t-t_1)}\right] \quad (t_1 \le t \le t_2) \tag{4.34}$$

Fig. 4.5.4 shows the variation of the time-dependent mass of the PM deposited in the DPF determined from the measured Δp in Fig. 4.5.2 and Eq. (4.29). The theoretical estimation curve obtained from Eqs. (4.32) and (4.34) is also plotted. The values of the parameters k', B, and C are determined by fitting the curve with the experimental data. The value of the rate constant is obtained as $k' = 3.0\ h^{-1}$ for both the seventh and the eighth trials. The other parameters for the seventh trial are $B = 1.91 \times 10^{29}\ m^{-3}\ h^{-1}$ and $C = 2.54 \times 10^{29}\ m^{-3}$, and for the eighth trial, they are $B = 2.55 \times 10^{29}\ m^{-3}\ h^{-1}$ and $C = 8.83 \times 10^{28}\ m^{-3}$. The theoretical estimation curve agrees well with the experimental data, which means that the characteristics of the DPF regeneration can be described with the present model.

4.6 Pilot-scale plasma particulate matter decomposition

4.6.1 Diesel particulate filter regeneration

Pilot-scale DPF regeneration using plasma-induced ozone injection for marine diesel engines is explained here [21]. On the basis of the principle of

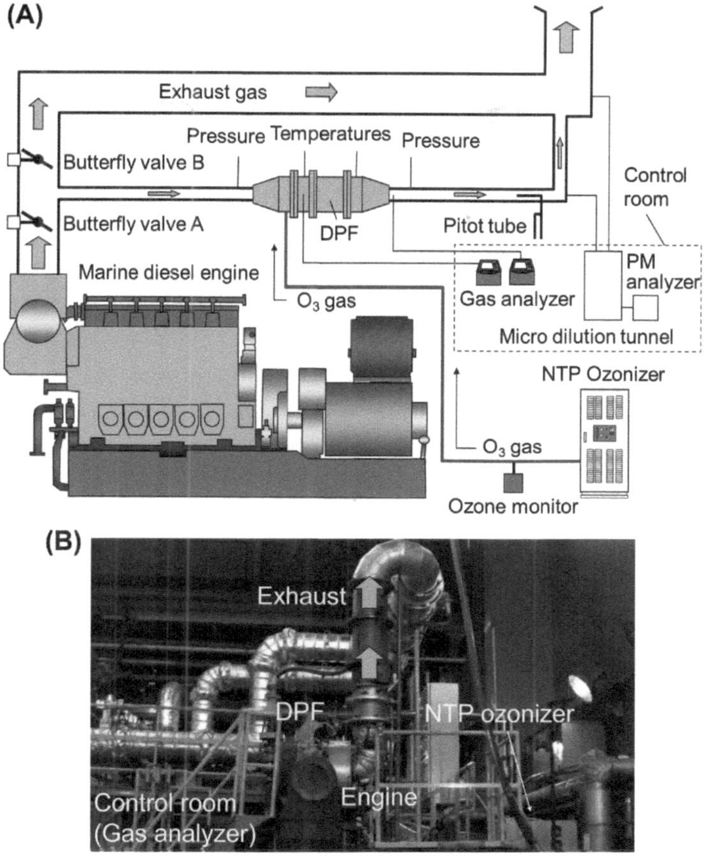

Figure 4.6.1 Experimental setup for diesel particulate filter (DPF) regeneration in marine diesel engine generator [21]. (A) Schematics of experimental setup and (B) photographic image of experimental setup.

DPF regeneration, a pilot-scale experiment is performed with a marine diesel engine. The technology is under the development of future promising one for marine diesel engines. Fig. 4.6.1A shows a schematic of the pilot-scale experimental apparatus and a photographic image thereof is shown in Fig. 4.6.1B. The engine that is tested is a marine diesel engine generator (5DK-20, Daihatsu Diesel MFG. Co., Ltd., Japan), which is used as an electrical power generator on ships such as an automobile carrier. The main specifications of the engine are as follows: a maximum output power of 800 kW, four cycles, five cylinders with a cylinder bore of 200 mm and a stroke of 300 mm, a constant rotation rate of 900 rpm, and marine diesel oil (A heavy oil, sulfur = 0.075 mass%, nitrogen = 0.01 mass%, heat quantity = 45.4 MJ/kg) as the fuel. The exhaust flow rate is 1460 Nm^3/h for a 25% load or output power and 2530 Nm^3/h for a 50% load or output power.

In the figure, the exhaust gas from the engine is divided into two flow paths—a flow path to the DPF and another to the bypass. Approximately 5% of the entire exhaust gas passes through the DPF. The flow rate of the bypassed gas is 50−120 Nm^3/h. The ceramic DPF (material: SiC, TYK Corporation, Japan) has the following specifications: a diameter of 190.5 mm, a length of 100 mm, a cell density of 169 cells/in^2, and a wall thickness of 16 mil (=16×10^{-3} in). More than 95% of the PM in the exhaust gas is captured in the DPF. The dimensions of the DPF are designed by taking into account the flow rate of the exhaust gas. The flow of the exhaust gas into the DPF is controlled by butterfly valves A and B, and its velocity is measured with a Pitot tube.

Fig. 4.6.2 shows the configuration of the DPF duct and the ozone injection tubes. One pair of ozone injection tubes, with six injection ports at intervals of 15 mm, is set before the exhaust gas passes through the DPF. To know the degree of the DPF regeneration by O_3 injection, the pressures of the gas before and after it passes through the DPF are measured using semiconductor-type pressure sensors (amplifier: AP-40RA, sensor: AP-44,

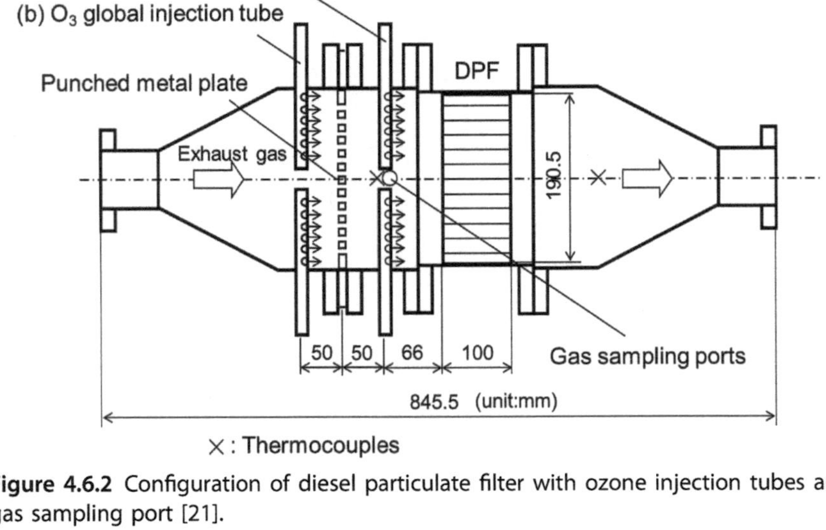

Figure 4.6.2 Configuration of diesel particulate filter with ozone injection tubes and gas sampling port [21].

KEYENCE Corporation, Japan). A data logger (NR-1000, KEYENCE Corporation, Japan) and thermocouples are used to record temperature data. The PM concentration is measured via the sampling ports by using a filtration method using PTFE microfiber-type filters with a diameter of 70 mm (the collection efficiency is 99.9% for dioctyl phthalate particles with a diameter of 0.3 μm); this measurement is based on the procedure described in ISO 8178. A microdilution tunnel apparatus (inner diameter of the dilution tunnel = 35.7 mm, MDLT-1300TM, Horiba, Ltd., Japan) is used for PM filtration. NO_x (NO and NO_2), CO, CO_2, and O_2 concentrations are measured by using a gas analyzer (PG-235, Horiba, Ltd., Japan) via the sampling port located upstream of the DPF. The concentrations of HCs compounds and SO_x are measured using gas detection tubes (No. 103 and 105 for HCs, and No. 5L for SO_2, Gastec Co., Japan). It should be noted that the concentrations of HCs are measured based on the octane rating. A punched metal plate is placed upstream of the DPF to unify the direction of the injected ozone flow. Ozone injection is carried out in two different modes: (1) global injection, in which O_3 is globally injected via two tubes upstream of the punched metal plate and (2) partial injection, in which O_3 is injected locally via two tubes downstream of the punched metal plate.

Table 4.6.1 lists detailed specifications of the ozonizer (EW-90Z, Ebara Jitsugyo Co., Ltd., O_3 = 90 g/h, input power consumption = 3.1 kW), which is equipped with a pressure swing absorption (PSA) oxygen generator that generates oxygen from ambient air. It should be noted that a silent discharge-type reactor is used. O_3 gas is generated after oxygen gas passes

Table 4.6.1 Specification of ozonizer.

O_2 production method from air	PSA
O_2 and O_3 flow rate	0.9 Nm^3/h
O_3 production method from O_2	Silent discharge
O_3 yield	0−90 g(O_3)/h
O_3 concentration	0−100 g(O_3)/Nm^3 = 0−4.67%
Power consumption	1.5 kW (PSA) 1.6 kW (discharge) 3.1 kW (total)
Dimensions	0.7 W × 0.92D × 1.8H (unit: m)
Discharge voltage	10 kV
Discharge current (max.)	5.5 A
Discharge frequency	0−6.82 kHz

PSA, pressure swing absorption.

through the plasma reactor. Thirty plasma reactors of this type are equipped with the ozonizer.

4.6.2 Results of regeneration

The measured PM emission rate per engine output power of the present marine engine generator, which is measured at the main exhaust gas flow, is 0.27 g/kWh (= 0.037 g/Nm^3) for a 25% engine load and 0.095 g/kWh (= 0.015 g/Nm^3) for a 50% engine load. Typical concentrations of the untreated gas are as follows: NO_x = 860–890 ppm (NO = 790–840 ppm, NO_2 = 40–70 ppm), CO = 120–340 ppm, HCs = 200 ppm, and SO_2 = 20 ppm for a 25% engine load, and NO_x = 940–1040 ppm (NO = 860–940 ppm, NO_2 = 70–110 ppm), CO = 110–140 ppm, HCs = 200 ppm, and SO_2 = 30–60 ppm for a 50% engine load. The concentrations fluctuate according to the variation in the engine operating parameters and of the exhaust gas temperature from 300 to 320°C. The residence times of O_3 between the injection tube and the end of the DPF are 0.16–0.30 s for a 25% engine load and 0.10–0.16 s for a 50% engine load.

Because the pressure difference also depends on the gas temperature, the resistance coefficient C_R is introduced, which is defined as follows:

$$C_R = \Delta P/Q \tag{4.35}$$

$$Q = Q_N T_{avg}/T_N \tag{4.36}$$

where ΔP is the pressure difference at the DPF, Q is the actual exhaust gas flow rate at the DPF, Q_N is the exhaust gas flow rate in the standard state (0°C, one-pressure), T_{avg} is the average of the two measured absolute temperatures, and T_N is the absolute temperature in the standard state (=273.15 K).

Fig. 4.6.3 shows an example of the measurement of the pressure difference and exhaust gas flow rates during DPF regeneration using global injection when the flow rate of the exhaust gas is 80 Nm^3/h and the temperature is 270–320°C. The resistance coefficient C_R decreases only when the plasma is turned on, and further, DPF regeneration clearly occurs in these cases. The gradient of the C_R curves indicates that an exhaust gas temperature of 320°C is beneficial for DPF regeneration.

The relationship between O_3 injection and the NO_x oxidation rate is investigated when the global injection is employed. The data are obtained under the following engine operating conditions: engine loads of 25%

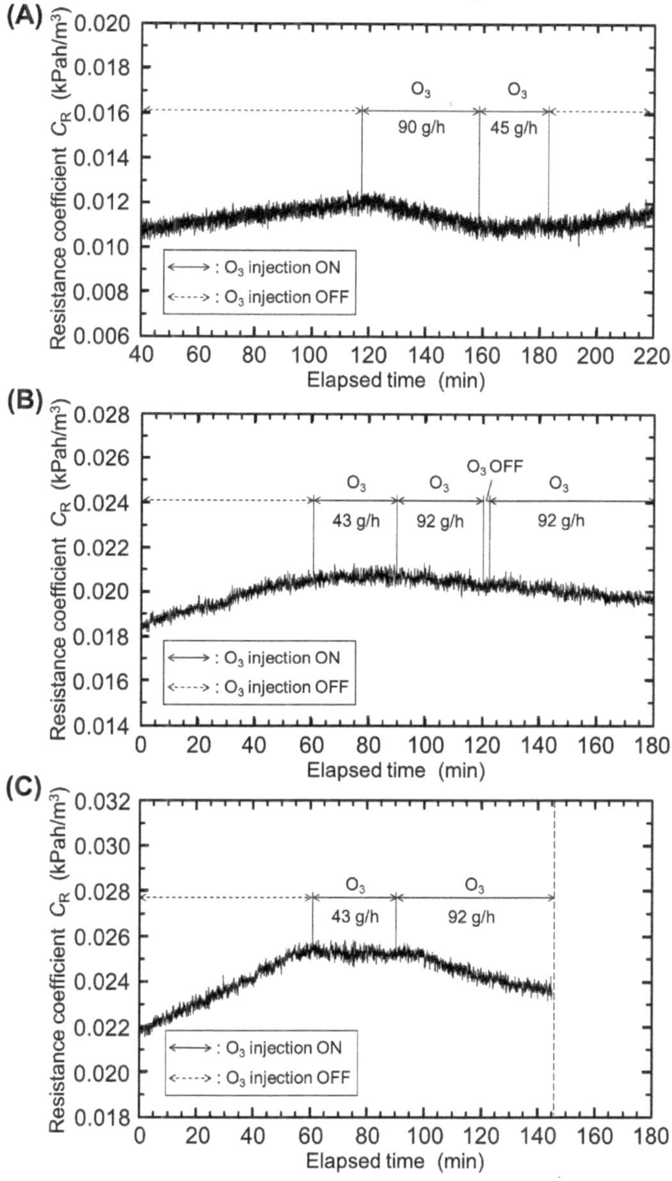

Figure 4.6.3 Diesel particulate filter (DPF) regeneration results at different temperatures for 80 Nm^3/h using global injection [21]. (A) Exhaust gas temperature of 270°C at the DPF and engine load of 50%, (B) exhaust gas temperature of 270°C at the DPF and engine load of 25%, and (C) exhaust gas temperature of 320°C at the DPF and engine load of 25%.

and 50%, exhaust gas flow rates of 50, 80, and 120 Nm^3/h, and exhaust gas temperatures of 220–320°C. In the discussion, the data are compared with the theoretical relationship that is obtained with the NO oxidation rate according to the following reactions:

$$O_3 \rightarrow O_2 + O \tag{4.37}$$

$$NO + O \rightarrow NO_2 \tag{4.38}$$

These species contribute to PM oxidation as

$$NO_2 \rightarrow NO + O \tag{4.39}$$

$$C + 2O \rightarrow CO_2 \tag{4.40}$$

$$C + O \rightarrow CO \tag{4.41}$$

Under any of these conditions, the oxidation rate of NO increases when the O_3 injection rate increases. The NO oxidation rates determined experimentally are smaller than those obtained by the theoretical relationship. The NO oxidation rates become lower with an increase in the flow rate and an increase in the exhaust gas temperature. This result could be due to the thermal loss of NO_2 of

$$2NO_2 \rightarrow 2NO + O_2 \tag{4.42}$$

and the loss of O_3 by other oxidation processes such as the following reactions:

$$SOF + lO \rightarrow mCO_2 + nH_2O \quad (l,\ m,\ \text{and } n \text{ are integers}) \tag{4.43}$$

$$2HC + 5O \rightarrow 2CO_2 + H_2O \tag{4.44}$$

where HC means hydrocarbon compound as one of the PMs. It is noted that reactions (4.37)–(4.41), (4.43), and (4.44) have already been explained in Subsection 3.5.3. PM oxidation is caused by oxygen radical O induced by direct reactions (4.37) with O_3 and by indirect reactions (4.38) and (4.39) with NO_2. In the present experiment, almost 70% of the NO can be oxidized when the maximum ozone injection rate is used (90 $g(O_3)/h$). In contrast to the direct reactions, the indirect reactions may require lower temperature and occur mainly on the surface of the DPF. The shape, size, and chemical composition of the PM under different test conditions could affect the oxidation rates. However, the present study does not produce any experimental data in this regard, and the effect of these parameters would need to be examined in the future.

Continuous DPF regeneration results are obtained for the two different modes of the O_3 injections, which are global and partial injections with O_3 injection flow rates of 43 and 92 g/h. The resulting data provide the time gradients of the resistance coefficient C_R in the time series. The results demonstrate that the decrease in C_R is greater in the case of partial injection, in which the O_3 injection tube is located downstream of the metal plate. However, the C_R increases more rapidly when O_3 injection is discontinued because the DPF can be partially regenerated.

Fig. 4.6.4 shows the resultant photographs of the DPF cross section at the inlet side: (A) before regeneration, (B) regeneration by global injection, and (C) regeneration by partial injection. The duration of DPF regeneration in all the cases is 300 min. The engine load is 25%, and the exhaust gas temperature is 300−320°C. The photographs show that, although both types of injections achieve DPF regeneration, this most strongly occurs in the area surrounding the outlet of the injection tube. The ozone consumption rates, which is the mass of ozone consumed per mass of PM oxidized, for global and partial injections are 28 g(O_3)/g(PM) and 20 g(O_3)/g(PM), respectively. The experiments in the two injection modes are performed under the same conditions—an engine load of 25%, an exhaust gas flow rate

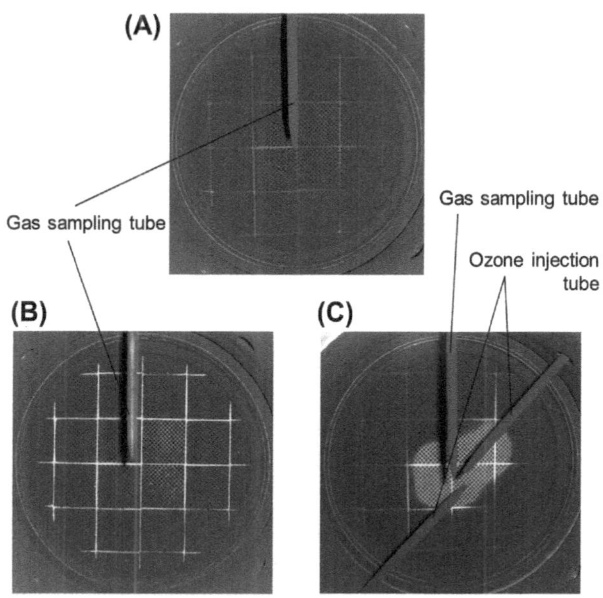

Figure 4.6.4 Photographs of diesel particulate filter regeneration results [21]. (A) Before regeneration, (B) after regeneration by global injection, and (C) after regeneration by partial injection.

of 50 Nm^3/h, and an exhaust gas temperature of 270°C. A smaller value for the ozone consumption rate indicates a higher degree of PM oxidation. The results show that the ozone consumption rate obtained with partial injection is smaller than that obtained with global injection, that is, partial injection with a high concentration of O_3 is more effective.

Throughout the tests, the energy efficiency of DPF regeneration attained maximum values of 2.7 g(PM)/kWh and 1.9 g(PM)/kWh in the case of 25% and 50% engine loads, respectively. On the basis of this value of 1.9 g(PM)/kWh, regarding the aforementioned PM emission rate per engine output power of 0.095 g(PM)/kWh for a 50% engine load, it is estimated that, for the present aftertreatment system, the plasma energy required for continuous DPF regeneration is approximately $(0.095/1.9) \times 100 = 5\%$ of the engine power output when the entire exhaust gas passes through the DPF.

4.7 Plasma NO_x reduction

4.7.1 Laboratory test on NO_x reduction of diesel engine exhaust gas

SCR technology is frequently applied to marine diesel engines for NO_x reduction. However, the SCR technology is problematic because a relatively high value of the exhaust gas temperature of 300°C is required for catalyst activation. To overcome this, the possibility of applying a plasma burner to improve the NO_x-reduced performance of a marine engine and to utilize it for catalyst regeneration was investigated by Jo et al. [22]. The apparatus of the plasma burner can enhance the SCR catalysts and increase the temperature. It was confirmed that decomposition of nitrogen oxide (De-NO_x) performance can be improved by increasing the temperature of the SCR with the plasma burner. Although this is a very effective technology, the other drawbacks exist such as the possible production of harmful nanoparticles and ammonia leaks, the use of harmful heavy-metal SCR catalysts, the necessary construction of urea solution supply stations, and degradation of the catalyst under the influence of sulfur in fuel. As an alternative technology, NO_x reduction with NTP has been proposed and developed. Laboratory-scale experiments are conducted against this background [23]. The engine that is used is a stationary diesel generator (YDG200VS-6E, Yanmar Co., Ltd., specification: single cylinder, four-stroke, rotation speed: 3600 rpm, maximum output: 2.0 kW). Marine diesel oil (MDO, sulfur = 0.067%) was used as the fuel.

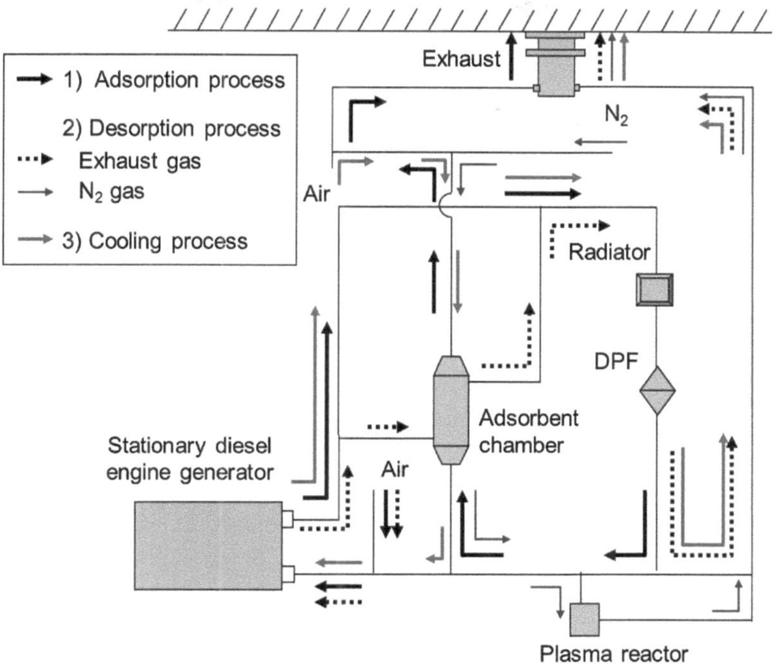

Figure 4.7.1 Outline of the device used in the laboratory experiments [23].

Fig. 4.7.1 shows a schematic diagram of the system. The system comprises a diesel engine, an adsorption chamber packed with copper oxide/manganese oxide pellets to adsorb the NO_x in the exhaust gas at a controlled temperature, an NTP reactor (surface discharge type, input power = 120 W), and a DPF. The exhaust gas flow rate is 300 NL/min. Exhaust gas aftertreatment includes 1) adsorption process followed by 2) desorption process and 3) cooling process. The flow directions of the exhaust gas in processes 1), 2), and 3) are indicated by the different arrows in the figure. In the adsorption process 1), the exhaust gas is cooled by an air-cooled radiator and passed through the DPF to remove PM. Then, the exhaust passes through the adsorption chamber, where NO_x is removed or adsorbed by the adsorbent such as zeolite pellets. The clean gas is discharged from the system. In the desorption process 2), the adsorption chamber is housed in a heat exchanger; exhaust gas passes through the first flow path, heating the adsorbent pellets packed inside the second flow path, through the walls of the heat exchanger, causing adsorbed NO_x to desorb. At the same time, nitrogen gas at a low flow rate (10 NL/min), which is used to provide low concentration oxygen gas, flows along the second flow path.

Consequently, $NO_x + N_2$ gas (the NO_x concentration is ordinarily approximately 2%) is formed by mixing with the desorbed NO_x. This $NO_x + N_2$ gas is then subjected to NTP, and high-efficiency reduction of NO_x is performed by employing the plasma reactor, yielding N_2 and O_2. In the cooling process 3), the remaining NO_x adsorbed to the pellets is desorbed by the introduction of air, at a flow rate of 50 NL/min, into the adsorption chamber with the aid of the remaining heat. The desorbed NO_x is recycled to the engine air intake. This recycled procedure is named exhaust gas component recirculation (EGCR). In the procedure, the remaining NO_x is broken down inside the engine, achieving the complete elimination of NO_x. It is noted that EGCR is an effective NO_x-reducing technology for diesel emissions because higher concentration NO_x is decomposed inside the engine. EGCR technology differs from exhaust gas recirculation (EGR) because higher concentration NO_x is recirculated with the adsorbents in EGCR, but lower concentration NO_x is directly recirculated without the adsorbents in EGCR.

Fig. 4.7.2 presents the experimental results and shows the time-dependent changes in the mass flow rate of NO_x discharged when untreated and when treated. The open circles show the mass flow rate of untreated NO_x, and the solid circles show the mass flow rate of treated NO_x. In the adsorption process in each cycle, the untreated NO_x discharged from the engine, indicated by open circles, decreases to the level of the solid circles. In contrast, the spikes in the mass flow rate of untreated NO_x in the desorption and cooling processes result from the addition of desorbed NO_x (flow path II) to the exhaust gas, which passes through the heat exchanger in

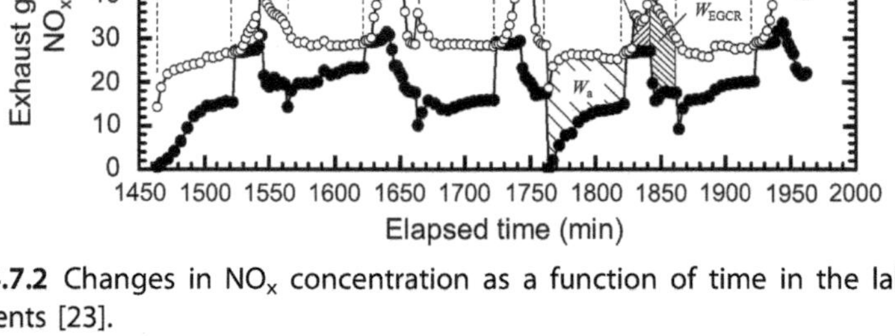

Figure 4.7.2 Changes in NO_x concentration as a function of time in the laboratory experiments [23].

flow path I without passing through the adsorption chamber and is then discharged to the outside. This means that the mass flow rate of NO_x decreases to the level of the solid circles owing to plasma treatment and EGCR.

The quantity of treated NO_x converted to NO_2 per unit of plasma energy (the energy efficiency) for the system is calculated, and 437 g(NO_2)/kWh is found to be the highest value. With normal plasma treatment, NO_x is at a low concentration (of the order of several hundred ppm), and therefore, the highest possible efficiency is, at most, approximately 20 g(NO_2)/kWh; however, the concentration of NO_x (to ca. 2%) and the effect of EGCR effectively achieve the high plasma energy efficiency reported here.

4.7.2 Pilot-scale test on NO_x reduction of marine diesel engine exhaust gas

Next, based on the laboratory-scale experiments, a pilot-scale device in the form of an actual marine diesel engine (output class 1 MW) was designed, and an aftertreatment system in the factory was developed [24]. However, because heavy oil, which contains sulfur, is used as fuel for the engine, the exhaust gas includes PM that is typically a sulfuric acid mist. Moreover, a turbocharger is used in the engine air intake system, and therefore, exhaust gas recycling could possibly damage the engine. In these experiments, EGCR explained in the previous subsection is not carried out, and the experiments were carried out by alternately repeating 1) adsorption and 2) desorption (plasma treatment).

The experiments are carried out using a marine diesel engine (for auxiliary power generator on board, 6DK-20e, Daihatsu Diesel MFG. Co., Ltd., specification: four-cycle, six cylinders, rotation speed: 900 rpm, maximum output: 1081 kW, exhaust gas flow rate: 6815 Nm^3/h for maximum output). The aftertreatment experimental system constitutes a scaled-up adsorption chamber, an NTP reactor with 48 connected surface discharge tubes (total output 21.4 kW), a DPF, and a cooler. DPF regeneration employing plasma-excited ozone O_3 is carried out in a parallel experiment, aiming for complete exhaust gas cleaning. Part of the exhaust gas from the engine is diverted, and the exhaust gas is treated at a flow rate of 800 Nm^3/h. Aftertreatment includes an adsorption and a desorption process, and these processes are repeatedly alternated. In the desorption process, N_2 gas at 12–24 m^3/h is introduced to the adsorption chamber in the desorption process. The $NO_x + N_2$ gas that flows out is treated by the surface discharge NTP reactor.

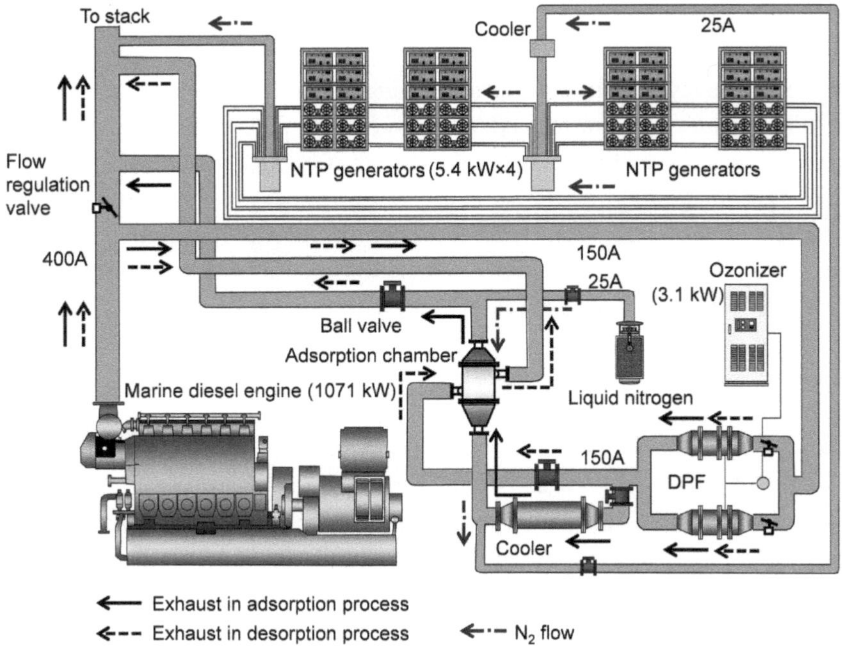

Figure 4.7.3 Experimental setup of the aftertreatment system with the marine diesel engine [24].

Fig. 4.7.3 shows the experimental setup for the aftertreatment system for the marine diesel engine. The exhaust gas emitted from the engine is first separated, and approximately 13% of the bypassed gas, of which the flow rate is controlled by a flow regulation valve, passes through a set of ceramic DPFs, where most PM is removed. After passing through the DPFs, the flow velocity of the gas is measured by a set of pitot tubes (L type, FV-21A, Okano Works, Ltd., Japan). The size of the pipe for the bypassed flow is 150 A (in Japanese Industrial Standards; inner diameter = 155.2 mm). The DPF (material: SiC, TYK Corporation, Japan) has the following specifications: a diameter of 190.5 mm, a length of 100 mm, a cell density of 169 cells/in^2, and a wall thickness of 16 mil (= 0.406 mm). The PM accumulated in the DPF is treated by NTP-induced ozone (O_3) injection technology, as previously described for a marine diesel engine. O_3 is generated by an ozonizer (EW-90Z, Ebara Jitsugyo Co., Ltd., Japan, O_3 = 90 g/h, input power = 3.1 kW) equipped with a PSA oxygen generator. After the removal of PM, the NO_x in the exhaust gas is treated by NTP combined with adsorption and desorption processes. The NO_x reduction process is realized by switching valves and controlling the flows of the adsorption and desorption processes. The exhaust gas is cooled from 270°C to approximately 45°C with a water cooler to ensure the effective adsorption of NO_x to the adsorbents. A maximum of 92% of the NO_x is adsorbed in the chamber containing the adsorbents. Before the gas passes through the chamber, the concentration of untreated NO_x arriving from the engine is 757−963 ppm, and the ratio of NO_2 to NO_x is approximately 10%. Finally, the exhaust gas exits the stack. In the desorption process, after the valves have been switched accordingly, N_2 gas at a lower flow rate of 11.1−22.2 Nm^3/h from a liquid N_2 cylinder passes through the chamber. The flow direction of the N_2 gas is opposite to that of the exhaust gas during adsorption. Simultaneously, the chamber is heated by waste heat from the exhaust gas. As a result, high-concentration (typically 4000 ppm) NO_x flows out of the chamber and enters the NTP reactors with a maximum total energy consumption of 21.6 kW. NO_x is reduced and purified according to the following reaction:

$$2NO_x + 2N \rightarrow 2N_2 + xO_2 \tag{4.45}$$

The concentrations of NO_x, NO, CO, and O_2 after adsorption and after NTP treatment are measured using a gaseous analyzer (PG-235, NO_x chemiluminescence-type, Horiba, Ltd., Japan). The N_2O concentration after NTP treatment is measured using a gaseous analyzer (VIA-510, infrared-type, Horiba, Ltd., Japan) through a sampling unit (ES-510, Horiba, Ltd., Japan) and a catalyzer unit (Horiba, Ltd., Japan). The total NO_x and O_2 concentrations before adsorption and after desorption are measured using another NO_x and O_2 analyzer (NOA-7000, Shimadzu Corporation, Japan). In the case of high-concentration desorbed NO_x exceeding 2500 ppm, which is the upper limitation of the analyzer, the gas is diluted with atmospheric air and the real NO_x concentration is estimated by comparing the oxygen concentration of the diluted exhaust gas with that of the raw exhaust.

In the figure, the adsorption chamber equipped with a waste heat exchanger specially designed and manufactured by Sumitomo Precision Products Co., Ltd. is used. In a cross section of the chamber, two types of flow paths exist, namely, flow path I and flow path II, alternately stacked inside. Flow path I (the number is 47 and each gap is 3.2 mm) is empty and flow path II (the number is 48 and each gap is 8.9 mm) is packed with adsorbent pellets to a height of 270 mm. The thickness of the plates between path I and II is 1 mm. Corrugated stainless steel heat exchanger fins are placed in flow path II to enhance heat transfer. In the adsorption process, exhaust gas flows from the bottom inlet to the top outlet of flow path II, and the NO_x is adsorbed onto the adsorbent pellets. In the desorption process, the heated exhaust gas passes through flow path I to heat the adsorbent pellets. Simultaneously, nitrogen gas or oxygen lean gas flows at a low flow rate from the top inlet to the bottom outlet of flow path II. The flow is switched between these two processes by changing the gas flow directions by opening and shutting the ball valves. The adsorbent used in this study is a MnO_x−CuO oxidative compound (N-140, 1.2−2.4 mm-sized granular pellets, Süd-Chemie Catalysts Japan, Inc.). The heat exchanger in the adsorption chamber is a counter-flow type, and the design specifications of the adsorption chamber are as follows: the mass of the chamber is 950 kg, amount of packed adsorbent pellet is 80 kg, the pressure drop is 2 kPa, and space velocity is 16,000 h^{-1}. The total heat exchange quantity, Q_{ex}, is estimated as 61.2 kW.

Fig. 4.7.4 shows the concentrations of raw or untreated exhaust gas components, i.e., NO_x, CO, CO_2, O_2, and HCs from the targeted diesel engine at engine load of 10−100%. The concentration of HC is shown with a unit of ppmC, which means the concentration C1-based HC determined from the number of carbon atoms or methane equivalent concentration. The data are measured as a specification of the engine itself using a marine diesel exhaust gas measuring system (MEXA-1600DS, Horiba, Ltd., Japan). All aftertreatment experiments are performed at engine loads of either 50% or 75%. As shown in the figure, at an engine load of 50%, the typical concentrations of NO_x, CO, CO_2, O_2, and HCs are 660 ppm, 33 ppm, 4.8% (=48,000 ppm), 14.4%, and 120 ppm,

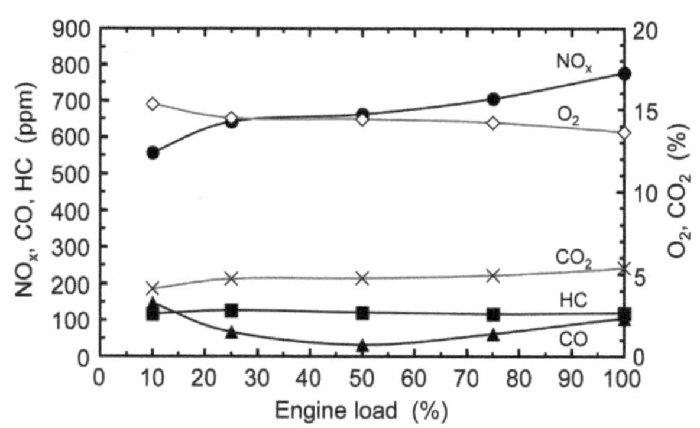

Figure 4.7.4 Exhaust gas components from the targeted diesel engine at engine loads of 10−100% [24].

respectively, and at an engine load of 75%, they are 710 ppm, 63 ppm, 4.9%, 14.2%, and 120 ppm, respectively.

Fig. 4.7.5 shows the NTP reactor used for NO_x reduction in the desorption process. The reactor consists of a surface discharge element (ET-OC70G-C, Masuda Research Inc., Japan), air-cooling fins, and a flange to fix the discharge element to the frame. Tungsten discharge electrodes are arranged on the inner ceramic wall (alumina) of the discharge element. Two of these reactors are powered by a single-pulse high-voltage (maximum peak-to-peak voltage = 10 kV, frequency = 10 kHz, and a maximum input power = 450 × 2 = 900 W) power supply (HCII-70/2, Masuda Research Inc.). A unit of NTP generators (HCII-OC70 × 12) consists of 12 NTP reactors and 6 power supplies. The total input power of a unit is 900 W × 6 = 5.4 kW and the discharge power is 5.0 kW. The electrical conversion efficiency is 5.0/5.4 × 100 = 93%. In the experiment, four NTP generator units (total input power is 5.4 × 4 = 21.6 kW) connected in three parallel separate flow lines are used, as shown in Fig. 4.7.5. The flow rate of each line is 3.7−7.4 Nm^3/h and the total is 11.1−22.2 Nm^3/h. When high-concentration desorbed NO_x in N_2 enters through the inlet of the reactor, the NO_x reduced by the NTP exists through the reactor outlet as a clean gaseous mixture of N_2 and O_2.

Fig. 4.7.6 shows the time-dependent NO_x emissions before and after the gas passes through the aftertreatment system for operating cycles 9−12. In this figure, the mass flow rate for NO_x shown on the vertical axis is

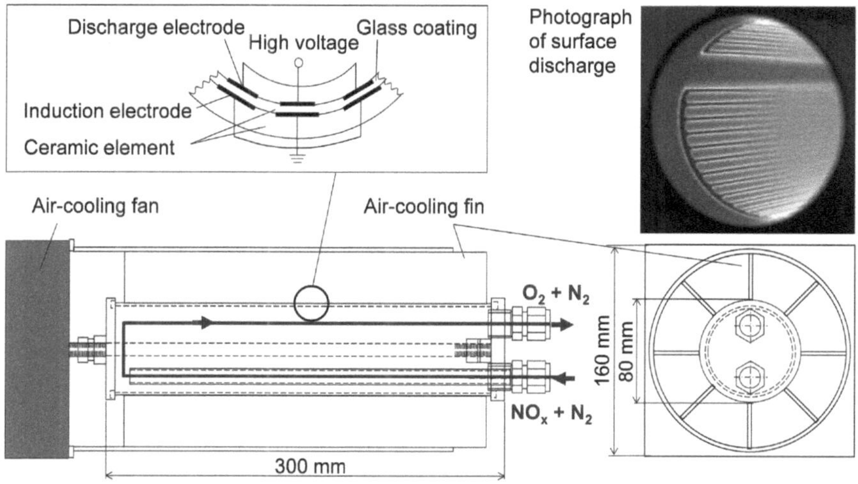

Figure 4.7.5 Schematic of the nonthermal plasma reactor, structure of discharge section, and photographic image of the inside of the reactor and surface discharge [24].

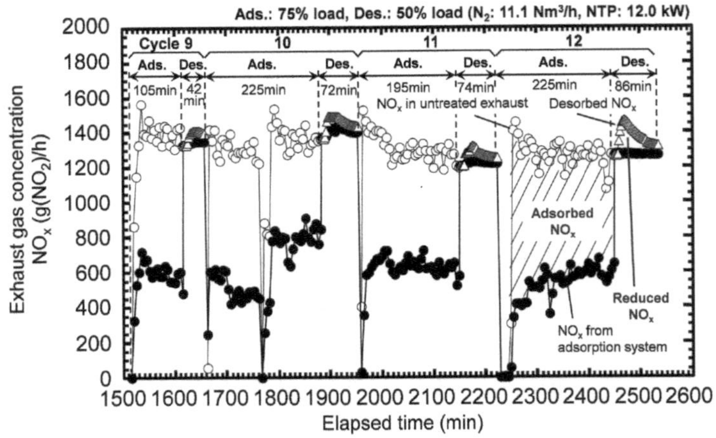

Figure 4.7.6 Time-dependent NO_x emissions before and after the gas passes through the aftertreatment system for operating cycles 9–12 [24].

evaluated based on the molecular mass of NO_2 with the unit of g(NO_2)/h. Furthermore, untreated NO_x is indicated by open circles with lines in the adsorption process, whereas in the desorption and cooling processes, it is indicated by open triangles with lines. NTP is applied to all of the desorption processes with a total input power of 12.0 kW. The NO_x mass flow rate of the untreated exhaust gas is approximately 1070–1570 g(NO_2)/h. As a result of the adsorption processes, the mass flow rate of NO_x decreases to 240–910 g(NO_2)/h. In the desorption processes, the characteristics of NO_x desorption is that the concentration drastically increases to reach a maximum. The maximum concentrations are 3460, 5200, 5030, and 8340 ppm in cycles 9–12. This high concentration in cycle 12 is also due to the larger amount of NO_x adsorbed in the adsorption process. Part of the NO_x is desorbed and reduced by the NTP generators, and 59% of NO_x is reduced on average. In cycle 12 in the plot, the hatched area bordered by open and solid symbols represents the total mass of adsorbed NO_x, and the area in the desorption process is the total mass of NO_x reduced by NTP.

Table 4.7.1 shows the adsorbed, desorbed, reduced, and treated amounts of NO_x together with the gas flow rates and energy efficiencies of NO_x removal by the NTP and the entire system in cycles 9–12. At cycle 10, the adsorbed mass of NO_x is $W_a = 2365$ g(NO_2), the desorbed mass of NO_x is $W_d = 109$ g(NO_2), the mass of NO_x reduced by the application of NTP is $W_{NTP} = 56.6$ g(NO_2), and the applied NTP energy is $E_{NTP} = 14.4$ kWh; therefore, the highest system energy efficiency of $\eta_{system} = 161$ g(NO_2)/kWh is recorded for this cycle. It is noted that in the adsorption process of cycle 10,

Table 4.7.1 Amount of NO_x treated and energy efficiency of removal in cycles 9–12 [24].

	Cycle	9	10	11	12
Averaged flow rate of exhaust gas	Nm^3/h	769	739	707	678
Adsorbed NO_x, W_a	$g(NO_2)$	1254	2365	2169	2412
Flow rate of N_2	Nm^3/h	11.1	11.1	11.1	11.1
Desorbed NO_x, W_d	$g(NO_2)$	41.2	109	89.3	159
Reduced NO_x by NTP, W_{NTP}	$g(NO_2)$	29.9	56.6	57.6	144
Removed NO_x in system, $W_{system} = W_a - W_d + W_{NTP}$	$g(NO_2)$	1243	2312	2137	2397
NTP power	kW	12.0	12.0	12.0	12.0
NTP energy, E_{NTP}	kWh	8.40	14.4	14.8	17.2
$\eta_{NTP} = W_{NTP}/E_{NTP}$	$g(NO_2)/kWh$	3.56	3.93	3.89	8.37
$\eta_{system} = W_{system}/E_{NTP}$	$g(NO_2)/kWh$	148	**161**	144	139

The highest system energy efficiency is shown as a bold numeral in cycle 10.

the typical concentrations of NO_2, NO, CO, and O_2 downstream of the adsorption chamber are 160 ppm, 270 ppm, 95 ppm, and 13.3%, respectively; in the desorption process of cycle 10, these concentrations downstream of the NTP generator are 460 ppm, 1570 ppm, 370 ppm, and 4.7%, respectively. In all cycles, the highest concentration of the desorbed NO_x is 7820 ppm, of which 71% is NO_2.

For the processes in Table 4.7.1, the time-dependent temperature of the adsorbent pellets in the adsorption chamber in cycles 9–12 is measured. The engine load is set to 75% for all adsorption processes and 50% for all desorption processes. Although the adsorbent temperature is higher than 250°C at the beginning of each adsorption process because of residual heat from the previous desorption process, the temperature rapidly decreases to 50°C owing to the cooled exhaust gas flowing into the chamber. The adsorption temperature is set higher in cycle 11 to determine the adsorption characteristics at higher temperature. Although the adsorption amount decreased slightly, the decrease is not notable in cycle 11. During the desorption process, the temperature rapidly increases due to heat exchange with the exhaust gas along flow path I. The efficiency of NO_x adsorption tends to improve at lower adsorbent temperatures. On the other hand, the efficiency of NO_x desorption tends to be enhanced at higher adsorbent temperatures. The aftertreatment system can be designed to achieve the appropriate temperatures for both the adsorption and desorption processes.

The IMO emission standards from Tier II to III for a marine diesel engine with a rotating speed of 900 rpm requires NO_x emissions to be reduced by 6.89 g/kWh, as explained in Chapter 2. The recorded system energy efficiency of $\eta_{system} = 161$ g(NO_2)/kWh in terms of the IMO emission standards from Tier II to III corresponds to only 4.3% (= 6.89/161×100) of the engine output power. This high-performance aftertreatment result fulfills the most recent IMO emission standards for 2016 using present-day technology.

In this experiment, because of the increasing cost to operate the engine, the number of repetitions is insufficient, and an equilibrium state could not be achieved, and therefore, the desorption of NO_x is lower than its adsorption. The maximum ratio of desorption to adsorption, W_d/W_a, is 6.6% in the final cycle 12. The present results show that NO_x adsorption mainly contributes to the high energy efficiency. However, the NTP process becomes more important when the amount of desorbed NO_x increases in the equilibrium state by repeating this aftertreatment. Repetition of the experiment until W_d/W_a approximates 100% and $W_a = 2412$ g(NO_2), which is the same as that in cycle 12, is considered to enable a system energy efficiency, η_{system} of 114 g(NO_2)/kWh to be obtained from a relationship $\eta_{NTP} = 0.0473\ W_d$ because η_{system} approaches η_{NTP} according to Eqs. (4.46)–(4.48).

$$W_{system} = W_a - W_d + W_{NTP} \tag{4.46}$$

$$\eta_{NTP} = \frac{W_{NTP}}{W_{NTP}} \tag{4.47}$$

$$\eta_{system} = \frac{W_{system}}{E_{NTP}} \tag{4.48}$$

Thus, the experiment should be continued and repeated to evaluate the performance of this aftertreatment system when the adsorbent is under steady-state conditions.

4.8 Exhaust gas recirculation method

4.8.1 Laboratory example of exhaust gas recirculation

The effect of EGR on a diesel engine is simulated experimentally with a gas add-on in an attempt to fundamentally understand the method which is reported in Ref. [13] in Chapter 3. EGR ratio can be given by a ratio of intake CO_2 concentration to exhaust CO_2 one. Fig. 4.8.1 schematically

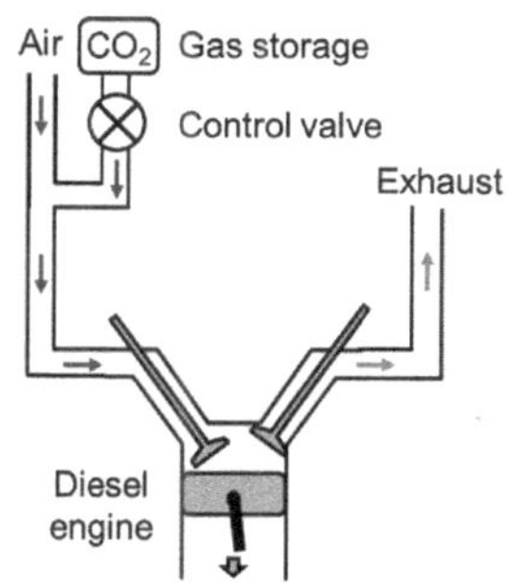

Figure 4.8.1 Schematic diagram of CO_2 gas add-on method in the experimental simulation of the exhaust gas recirculation effect.

illustrates the CO_2 gas add-on method in the experimental simulation of the EGR effect. This technology can be applied to two-cycle and four-cycle engines for both cases of automobile and ship. In the figure, the target is four-cycle engine. The experiment involves the synthesis of the intake gas mixture with air and/or pure gases stored externally. The influence of EGR is investigated by using CO_2 from an external storage source such as a gas cylinder by regulating the CO_2 concentration in an EGR-like intake mixture. The molar concentrations of O_2 and N_2 of the intake mixture are inversely proportional to the molar concentration of CO_2. This CO_2 add-on method simulates the thermodynamic and dilution effects of EGR. The compression temperature in a cylinder is governed by the quasiadiabatic compression process. As a result, in a diesel engine (displacement = 1.0 L, three cylinders, rotation speed = 1800 rpm), as the intake CO_2 increases at an add-on range of 0−19%, the compression pressure and the temperature in a cylinder decrease. For instance, the compression pressure decreases from 8.2 to 4.5 MPa, as the CO_2 concentration increases from 0% to 19%. In particular, it is observed dominantly in the range of the crank angles, i.e., 360–380 degrees where 360 degrees corresponds to top dead center. The decrease in the compression pressure and the temperature in a cylinder could contribute to the reduction of NO_x emission. The effectiveness of NO_x reduction by EGR varies with load. The heat capacity of the cylinder charge increases with the increase in CO_2 that is brought in by EGR. The flame temperature and, thus, the maximum temperature of the working fluid will be lowered with the increase in CO_2. Test results indicated that high ratios of EGR need to be applied at low load, but low ratios of EGR are sufficient for high load. This is because, when operating at lower loads, diesel engines generally tolerate a higher EGR ratio because the exhaust contains a high concentration of O_2 and low concentrations of combustion products CO_2 and H_2O. However, at high loads, the oxygen in the exhaust gas is insufficient, and inert constituents become dominating. In this study, hot raw EGR, in which hot exhaust is directly recirculated, and enhanced cooled EGR, in which cooled exhaust is recirculated, are investigated. The concentration of NO_x in the exhaust gas is typically 850 ppm without EGR. The hot raw EGR and enhanced cooled EGR result in NO_x concentrations of typically 680 and 600 ppm at an EGR ratio of 15%, respectively. The resulting NO_x concentrations in the hot raw EGR and enhanced cooled EGR are typically 600 and 450 ppm at an EGR ratio of 20%, respectively.

4.8.2 Pilot-scale example of exhaust gas recirculation in marine diesel engines

EGR systems are classified into low- and high-pressure exhaust gas recirculation (LP EGR and HP EGR) systems [25]. These EGR technologies can be applied to two-cycle and four cycle engines. In the subsection the case of two-cycle is explained. In general, for marine diesel engines, an EGR scrubber, a demister, and a water treatment system for PM and SO_x reduction in recirculated gas should be integrated into the EGR system because the engine is damaged by introducing uncleaned EGR gas. The difference of the position at which exhaust gas is taken in for recirculation exists for LP EGR and HP EGR. Fig. 4.8.2 shows a principle of a schematic diagram of the LP EGR system for marine diesel engines with an EGR scrubber, a demister, and a water treatment system. The water treatment system consists of collecting tank and centrifugal separators. In the EGR scrubber, which is a combination of a Venturi type and a packed tower type, SO_x is reduced by caustic soda (NaOH solution), and PM is also removed. NaOH solution is supplied to the EGR scrubber. The mist is removed by the demister and returned to the collecting tank. The solution in the collecting tank is cleaned by a pair of centrifugal separators. pH of solution is regulated. In the LP EGR system, the EGR gas exits from the outlet of the turbine of the turbocharger and enters the inlet of the compressor of the turbocharger as controlled by the EGR valve. Fig. 4.8.3 shows a principle of a schematic diagram of the HP EGR system for marine diesel engines with an EGR scrubber, a demister, and a water treatment system. Although the structure is almost the same as that in Fig. 4.8.2, in the HP EGR, the EGR gas enters the inlet of the turbine of the turbocharger and flows into the outlet of the compressor of the turbocharger as controlled by the EGR valve.

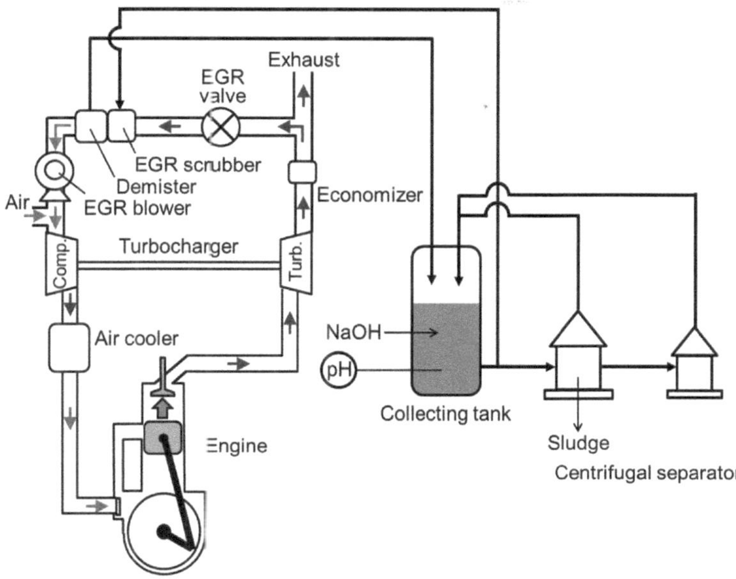

Figure 4.8.2 Schematic diagram of low-pressure exhaust gas recirculation (LP EGR) system with an EGR scrubber, a demister, and a water treatment system for marine diesel engines [25].

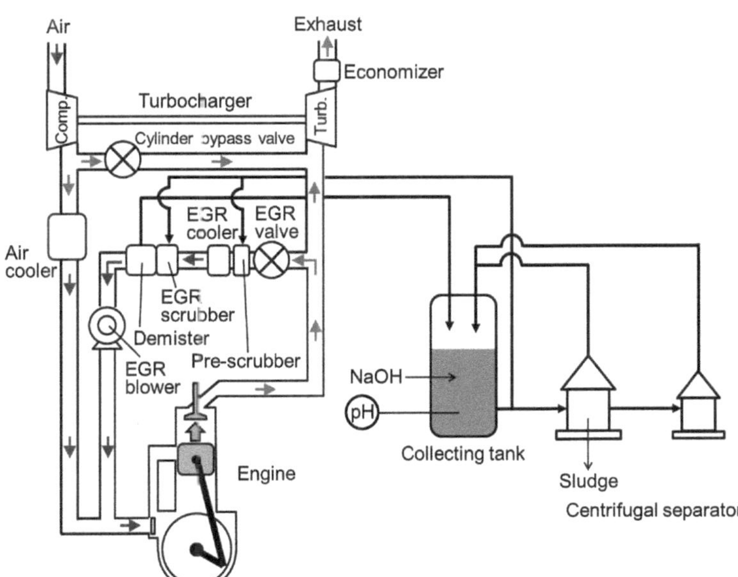

Figure 4.8.3 Schematic diagram of high-pressure exhaust gas recirculation (HP EGR) system with an EGR scrubber, a demister, and a water treatment system for marine diesel engines [25].

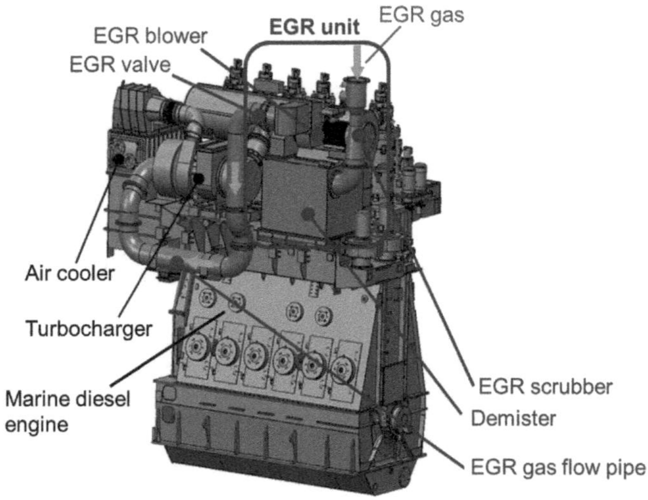

Figure 4.8.4 Schematic diagram of the low-pressure exhaust gas recirculation (LP EGR) system for marine diesel engine [25].

A recent example of the use of the LP EGR system in a marine diesel engine is Mitsubishi LP EGR system developed by Mitsubishi Heavy Industries, Ltd. and Japan Engine Corporation [26,27]. Fig. 4.8.4 shows a schematic diagram of the system [25]. The system can achieve NO_x reduction with LP EGR and also PM and SO_x removal from the exhaust gas with an EGR scrubber installed downstream of the EGR valve. A demister is installed between the EGR scrubber and the EGR blower. The scrubber is a combination of a Venturi type and a packed tower type scrubber [26]. Typically, the Venturi part cleans PM away, and the packed tower part removes SO_x. Furthermore, the water used in the EGR scrubber is recycled with a water treatment system, in which water from the acid scrubber is collected in a collecting tank after spraying into the exhaust gas, where the water is neutralized by controlling the pH value. The pH control is performed by using a pH sensor and NaOH solution (caustic soda). The PM in the neutralized water is removed by centrifugation leaving sludge as the residue. Treated water is obtained from these processes and is recycled. Miniaturization of the system is achieved, and the performance in engine operations with a single-cylinder test engine (NC33; two-cycle, bore diameter of 330 mm, Mitsubishi Heavy Industries, Ltd.) and a full-scale marine diesel engine (4UE-X3; two-cycle, bore diameter of 600 mm, output power of 9970 kW, Mitsubishi Heavy Industries, Ltd.) was investigated. An electronic control system, the Ecosystem, is introduced to optimize the timing of the fuel injection, exhaust valve, etc., during the evaluation of the system performance.

The result of the scrubber performance test showed that, in the case of an engine load of 100%, the SO_x concentration is 135 ppm (SO_2) at the scrubber inlet and 2.8 ppm (SO_2) at the outlet, that is, SO_x reduction of 98% is achieved. In the case of an engine load of 75%, the SO_x concentrations are 130 ppm (SO_2) at the scrubber inlet and 1.0 ppm (SO_2) at the outlet, that is, SO_x reduction of 99% is achieved. Further, they investigated soot removal and the SO_x rates against engine loads. The resulting soot removal rates are 81% at a load of 25%, 91% at a load of 50%, 91% at a load of 75%, and 88% at a load of 100%. The resulting SO_x removal rates are 91% at a load of 25%, 96% at a load of 50%, 99% at a load of 75%, and 96% at a load of 100%. Overall, typically SO_x removal exceeding 95% is achieved.

The LP EGR system with an EGR scrubber can reduce the O_2 concentration in the combustion chambers and add inert gases such as H_2O and CO_2 to the exhaust gas to increase the heat capacity. Therefore, the temperature of the combustion gas can be suppressed and NO_x generation is reduced. As a result of the NO_x reduction performance in EGR, in the case of an engine load of 75%, an increase in the EGR rate led to a gradual decrease in NO_x emission. NO_x reduction of approximately 80% is achieved at an EGR ratio of 35%. Further, they investigated the NO_x reduction rates against the EGR rates. For example, at an engine load of 100%, the resulting NO_x reduction rates are approximately 18% at an EGR rate of 5%, 30% at an EGR rate of 10%, 52% at an EGR rate of 20%, and 80% at an EGR rate of 30%. The deterioration rate of specific fuel oil consumption can be suppressed to within 1.5% by introducing this system.

As a recent example of the use of the HP EGR system, Mitsui Engineering and Shipbuilding Co., Ltd. developed the first unit of its large low-speed diesel engines with a low sulfur fuel-type HP EGR system [28]. Fig. 4.8.5 shows a schematic diagram of the system. The system can satisfy the Tier III NO_x regulation and is installed in a product tanker for the first time in Japan. For this class of marine diesel engines, the Tier III NO_x regulation requires NO_x emission up to 3.4 g/kWh, which corresponds to an 80% reduction compared with the Tier I regulation of NO_x emission up to 17.0 g/kWh and 76% reduction compared with the Tier II regulation of NO_x emission up to 14.4 g/kWh. The EGR devices are integrated into the diesel engine. As a result, a compact system is achieved. Therefore, the system has a lower impact on engine room design among technologies for

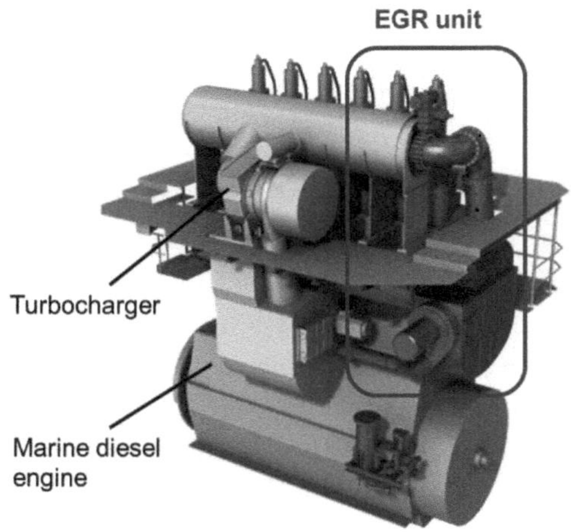

Figure 4.8.5 Schematic diagram of the high-pressure exhaust gas recirculation (HP EGR) system for marine diesel engine [28].

responding to NO_x regulations. It is announced in 2016 that this marine diesel engine with an HP EGR system (6G60ME-C9-EGRBP, Mitsui-MAN B&W) will be installed in two oil product tankers. The HP EGR system reduces NO_x in emissions by recirculating part of the exhaust gas from the engine to the scavenging pipe.

References

[1] G. Qi, R.T. Yang, Low-temperature selective catalytic reduction of NO with NH_3 over iron and manganese oxides supported on titania, Appl. Catal. B Environ. 44 (3) (2003) 217–225.

[2] Y. Kawada, S. Sato, S. Sato, M. Hosoda, A. Mizuno, Effect of non-thermal plasma on low temperature NO_x conversion with SCR catalyst, J. Inst. Electrostat. Jpn. 34 (5) (2010) 243–248 (in Japanese).

[3] H.-H. Kim, Y. Teramoto, A. Ogata, H. Takagi, T. Nanba, Plasma catalysis for environmental treatment and energy applications, Plasma Chem. Plasma Process. 36 (2016) 45–72.

[4] K. Hirata, N. Masaki, H. Ueno, H. Akagawa, Development of urea-SCR system for a heavy-duty commercial vehicles, SAE Technical Paper 2005-01-1860, 2005.

[5] K. Hanamoto, K. Goi, H. Tanimoto, SCR system complied with IMO NO_x tier III, J. Jpn. Inst. Mar. Eng. 51 (3) (2016) 269–272 (in Japanese).

[6] H. Nakaguchi, H. Tanaka, T. Nakao, Marine SCR system for large 2-stroke diesel engines, J. Jpn. Inst. Mar. Eng. 50 (5) (2015) 664–670 (in Japanese).

[7] M. Murakami, T. Nakao, Test results of marine SCR system for large diesel engines during long term service test, J. Jpn. Inst. Mar. Eng. 50 (1) (2015) 107–112 (in Japanese).

[8] Y. Watanabe, S. Koyanagi, Development and installation of marine-use hybrid SO_x scrubber system that compiles with IMO SO_x emission regulations, Mitsubishi Heavy Ind. Tech. Rev. 53 (2) (2016) 48–52.
[9] S. An, O. Nishida, New application of seawater and electrolyzed seawater in air pollution control of marine diesel engine, JSME Int. J. 46 (1) (2003) 206–213.
[10] T. Iuni, M. Yoshida, K. Koizumi, H. Kawakami, PM reduction technology with electrostatic precipitator for marine use, J. Jpn. Inst. Mar. Eng. 48 (4) (2013) 534–539 (in Japanese).
[11] S. Kambara, M. Okubo, et al., (Joint authorship), Taikiatsu Purazuma Kougaku Handbook–Hannou Katei no Kiso to Simulation no Jittsusai (Atmospheric Pressure Plasma Reaction Engineering Handbook–Fundamentals of Reaction Process and Actual Simulation), NTS, Inc., 2013, pp. 59–85 (in Japanese).
[12] M. Okubo, H. Fujishima, Fundamentals of electrostatic precipitator two-phase flow numerical simulation, J. Inst. Electrost. Jpn. 40 (4) (2016) 162–167 (in Japanese).
[13] T. Yamamoto, Y. Morita, H. Fujishima, M. Okubo, Three-dimensional EHD simulation for point corona electrostatic precipitator based on laminar and turbulent models, J. Electrost. 64 (2006) 628–633.
[14] T. Yamamoto, M. Okuda, M. Okubo, Three-dimensional ionic wind and electrohydrodynamics of tuft/point corona electrostatic precipitator, IEEE Trans. Ind. Appl. 39 (6) (2003) 1602–1607.
[15] Y. Li, C. Zheng, K. Luo, X. Gao, J. Fan, K. Cen, CFD simulation of high-temperature effect on EHD characteristics in a wire-plate electrostatic precipitator, Chin. J. Chem. Eng. 23 (2015) 633–640.
[16] Q. Liu, S.-S. Zhang, J.-P. Chen, Numerical analysis of charged particle collection in wire-plate ESP, J. Electrost. 74 (2015) 56–65.
[17] K. Kawakami, A. Zukeran, K. Yasumoto, T. Inui, Y. Enami, Y. Ehara, T. Yamamoto, Re-scattered particle trajectory simulation with a double cylinder type electrostatic precipitator, J. Jpn. Inst. Mar. Eng. 46 (5) (2011) 111–117 (in Japanese).
[18] H. Kawakami, A. Zukeran, K. Yasumoto, T. Inui, Y. Ehara, T. Yamamoto, Numerical simulation of three-dimensional particle migration and electrohydrodynamics of double cylinder electrostatic precipitator, Int. J. Plasma Env. Sci. Tech. 6 (2) (2012) 104–110.
[19] The Institute of Electrostatics Japan, Seidenki Handbook (Electrostatics Handbook), Ohmsha, Ltd., 1985, p. 47, p. 1204, p. 1125 (in Japanese).
[20] T. Adachi, T. Okubo, Ion wind generated by corona discharge, J. Inst. Electrost. Jpn. 11 (4) (1987) 246–254.
[21] T. Kuwahara, K. Yoshida, K. Hanamoto, K. Sato, T. Kuroki, T. Yamamoto, M. Okubo, Pilot-scale experiments of continuous regeneration of ceramic diesel particulate filter in marine diesel engine using nonthermal plasma-induced radicals, IEEE Trans. Ind. Appl. 48 (5) (2012) 1649–1656.
[22] S. Jo, D. Cho, D.H. Lee, K.-T. Kim, J.-O. Lee, Y.-H. Song, J.H. Jang, H.H. Roh, G.-M. Bak, T.-W. Lee, H.-S. Han, Plasma burner verification for SCR performance improvement and catalyst regeneration for marine engines, Trans. Korean Soc. Mech. Eng. B 42 (4) (2018) 305–313 (in Korean).
[23] T. Kuwahara, M. Nishimoto, K. Yoshida, T. Kuroki, M. Okubo, Development of plasma-assisted NO_x aftertreatment system for clean diesel engine (experiments with A type of heavy oil fuel), J. Jpn. Soc. Appl. Electromagn. Mech. 20 (1) (2012) 21–27 (in Japanese).
[24] T. Kuwahara, K. Yoshida, T. Kuroki, K. Hanamoto, K. Sato, M. Okubo, Pilot-scale aftertreatment using nonthermal plasma reduction of adsorbed NO_x in marine diesel-engine exhaust gas, Plasma Chem. Plasma Process. 34 (1) (2013) 65–81.

[25] N. Hiraoka, A. Miyanagi, K. Kurota, K. Ito, T. Nakagawa, T. Ueda, The world's first onboard verification test of UE engine with low pressure EGR complied with IMO's NO_x tier III regulations, Mitsubishi Heavy Ind. Tech. Rev. 53 (2) (2016) 40–47.
[26] M. Okabe, K. Sakaguchi, M. Sugihara, A. Miyanagi, N. Hiraoka, S. Murata, World's largest marine 2-stroke diesel test engine, the 4UE-X3 –development in compliance with the next version of environmental regulations and gas engine technology, Mitsubishi Heavy Ind. Tech. Rev. 50 (1) (2013) 55–62.
[27] T. Ueda, Development of low pressure exhaust gas recirculation system for exhaust gas emission control of Mitsubishi UE diesel engine, J. Jpn. Inst. Mar. Eng. 50 (6) (2015) 727–730 (in Japanese).
[28] Eureka International Inc, First MES EGR marine diesel fitted to product tanker—compliant with IMO tier III NO_x regulations, SEA-Jpn. (382) (2017) 3.

CHAPTER 5

Prospects for marine diesel engine emission control

Contents

New Technologies for Emission Control in Marine Diesel Engines
ISBN 978-0-12-812307-2
https://doi.org/10.1016/B978-0-12-812307-2.00005-5

5.1 Advanced NO_x control technologies

5.1.1 Marine dual-fuel engine

Maritime environmental issues are challenging to overcome. The International Maritime Organization (IMO) has imposed regulations on emission from marine diesel engines. The regulations are known as the IMO emission standards, which are referred to as IMO Tier I–III on NO_x emission control. To fulfill the standards, NO_x reduction technologies have been developed, and currently, improvements in engine combustion and selective catalytic reduction (SCR) technology are applied to marine diesel engines. However, to satisfy the future stringent regulations, a compact and advanced technology for the installation is required. In addition, the transportation of aqueous urea is a problem in SCR technology. Dual-fuel engines are expected to satisfy the requirement. A gas engine that uses natural gas as fuel is a very clean engine that emits relatively low NO_x, particulate matter (PM), CO_2, and zero SO_x. As an example, 43% of NO_x, 25% of CO_2, and 100% of SO_x can be reduced in comparison with oil [1]. Therefore, dual-fuel engines have many advantages and have been developed. A four-cycle dual-fuel engine is first developed in the late 1990s. A practical two-cycle dual-fuel engine appeared in 2010 [2]. Although the efficacy of SCR and exhaust gas recirculation (EGR) technologies are limited to NO_x reduction, dual-fuel engines can reduce not only NO_x but also SO_x, PM, and CO_2. In diesel engines, combustion is caused by the self-ignition of diesel fuel. However, in natural gas fuel, an ignition source is necessary for combustion. Different means for ignition are available. One of those is using an electrical discharge spark plug. In this method, a mixture of natural gas and air is ignited by the spark plug. Another method is using self-ignition. In this method, a mixture of natural gas and air is ignited by the self-ignition of sprayed diesel oil. Furthermore, there are methods that use a precombustion chamber as well. In summary, four different ignition methods exist in dual-fuel engines: ignition with the spark plug in the combustion chamber, ignition with the self-ignition of sprayed diesel oil in the combustion chamber, ignition with the spark plug in the precombustion chamber, and ignition with the self-ignition of sprayed diesel oil in the precombustion chamber.

In terms of the technical aspects of natural gas fuel engines, the problem of engine knocking should be overcome. Knocking is a state that the local self-ignition takes place at high pressures before flame propagation in a combustion chamber, and it leads to a phenomenon involving sharp sounds and vibration of the engine consequently. Knocking causes damage to engine materials owing to its shock waves. In engines, the compression ratio is defined as the ratio of the smallest to the largest volume of the combustion chamber during piston movement. To prevent knocking, two possibilities to resolve it exist: the compression ratio is lowered or the excess air ratio (air-to-fuel ratio) of the air and fuel mixture is increased to lower the temperature in the chamber. In practical dual-fuel engines, knocking is prevented by a sensitive control of the excess air-to-fuel ratio. However, some problems and difficulties remain. For instance, when a property change occurs in natural gas, natural gas having a low methane number causes knocking easily. The methane number is defined as an index value of the knocking resistance. The methane number of a fuel is given by considering knocking tendency of a fuel equivalent to that of methane and hydrogen mixture. Methane, which has a high knock resistance, is given an index value of 100. Hydrogen, which burns quickly relative to methane, has a low knock resistance and is given the index value of 0. If a gas mixture fuel has a methane number of 80, its knock resistance is equivalent to that of a gas comprising 80% methane and 20% hydrogen. The methane number of natural gas differs based on its origins. For instance, the methane number of natural gas from Scandinavia is almost 100, which is equivalent to methane purity of 100%. However, the number from Japan is typically 65 because the gas includes butane and propane. For natural gas that is supplied to a ship as liquefied natural gas (LNG), its methane number decreases when the LNG evaporates as gas for the engine fuel. It is noted that biogas often has a methane number more than 100.

The choice of engine lubricating oil is also difficult because the properties of diesel and gas engines fuels are different. In general, a lubricating oil having a high total base number (TBN) of 30–40 mg(KOH)/g is used for the marine fuel oil whose sulfur content is usually 3.5%. TBN is defined as a mass in milligrams of potassium hydroxide (KOH) equivalent to that of hydrochloric acid or perchloric acid to neutralize the total base in 1 mg of sample. It is noted that the unit of mg(KOH)/g means the equivalent number of milligrams of KOH per gram of oil sample. TBN stands for an alkalinity value that indicates the neutralization ability of an acid component, such as sulfuric acid, caused by sulfur in fuel for preventing damage to engine parts due to the corrosion of metal surfaces. A lubricating oil having a low TBN of ~5 mg(KOH)/g is used for a natural gas fuel whose sulfur content is 0%. In the use of a lubricating oil having a high TBN in natural gas fuel engine operation, combustion residues of alkaline substance deposit on the cylinder walls. As a result, the combustion residues, which have a heat-insulating effect, increase the combustion temperature, and the risk of knocking increases [2]. In actual operation, knocking is overcome by output power control. A dual-fuel engine can switch operations inside, i.e., diesel and gas modes. In the diesel mode, the fuel is diesel fuel. However, in the gas mode, 95–99% of the fuel is natural gas and the other is marine diesel oil (MDO) in terms of calorific value. In other words, natural gas is typically used, but diesel fuel is also used for the ignition. However, to enhance the performances of both diesel and gas engines, proper lubricating oils for the dual-fuel engine should be developed.

In the development of a marine dual-fuel engine, the technical points such as ignition and knocking problem with fuel and lubricating oil should be considered. From these viewpoints, a marine dual-fuel engine is explained according to Refs. [1,3]. Fig. 5.1.1 shows a photograph of a marine dual-fuel engine (6MG28AHX-DF, Niigata Power Systems Co., Ltd., Japan) [1]. Fig. 5.1.2 shows the schematic of the (A) diesel mode and (B) gas mode in the dual-fuel engine. The combustion system is a direct injection micropilot oil lean-burn system in the gas mode [3]. The rated rotational speed is 800 rpm. The numbers of cylinders for the corresponding rated outputs are six cylinders for 1920 kW, eight cylinders for 2560 kW, and nine cylinders for 2880 kW. Gas fuel is vaporized LNG, and liquid fuel is a MDO. In the diesel mode, shown in Fig. 5.1.2A, the engine operates as a conventional diesel engine, and its brake mean effective pressure is 2.0 MPa. This dual-fuel engine could be the first marine engine in Japan that can fulfill the IMO emission standards on NO_x Tier III regulation. This dual-fuel engine is used as a propulsion system for a tugboat. A problem in gas engines is the excessively small air-to-fuel ratio that causes engine knocking. Nevertheless, an excessively large air-to-fuel ratio causes misfire problem. For instance, in the gas mode, shown in Fig. 5.1.2B,

Figure 5.1.1 Photograph of marine dual-fuel engine. *(6MG28AHX-DF, in Niigata Marine Selection Guide, by courtesy of Niigata Power Systems Co., Ltd.)*

appropriate control of the air-to-fuel ratio is needed. In particular, although the engine output power rapidly increases, the corresponding increase in gas fuel makes the air-to-fuel ratio decrease and causes knocking. A diesel fuel injection valve, which is the conventional one, is installed in the cylinder head for the diesel-mode operation. In addition, a direct micropilot fuel oil injector is installed in the cylinder head for the ignition in the gas-mode operation. The injectors are mounted on a common rail. This system can maintain the same reliability in the diesel mode as the conventional diesel operation and achieves low NO_x emission with stability in the gas mode. A mode can be selected at any engine output power using a switching system between diesel and gas modes. When abnormal combustion is detected in the gas mode, the switching system shifts the operation instantly to the diesel mode. For the optimization of the air-to-fuel ratio, the temperature and pressure of the intake air into the engine is controlled with devices. This antiknocking technology with an abnormal combustion sensing device and operation mode switching system realizes a stable engine operation for a wide range of engine output powers. The problem of transient response, which emerges in conventional gas engines, is improved for applications in marine ships.

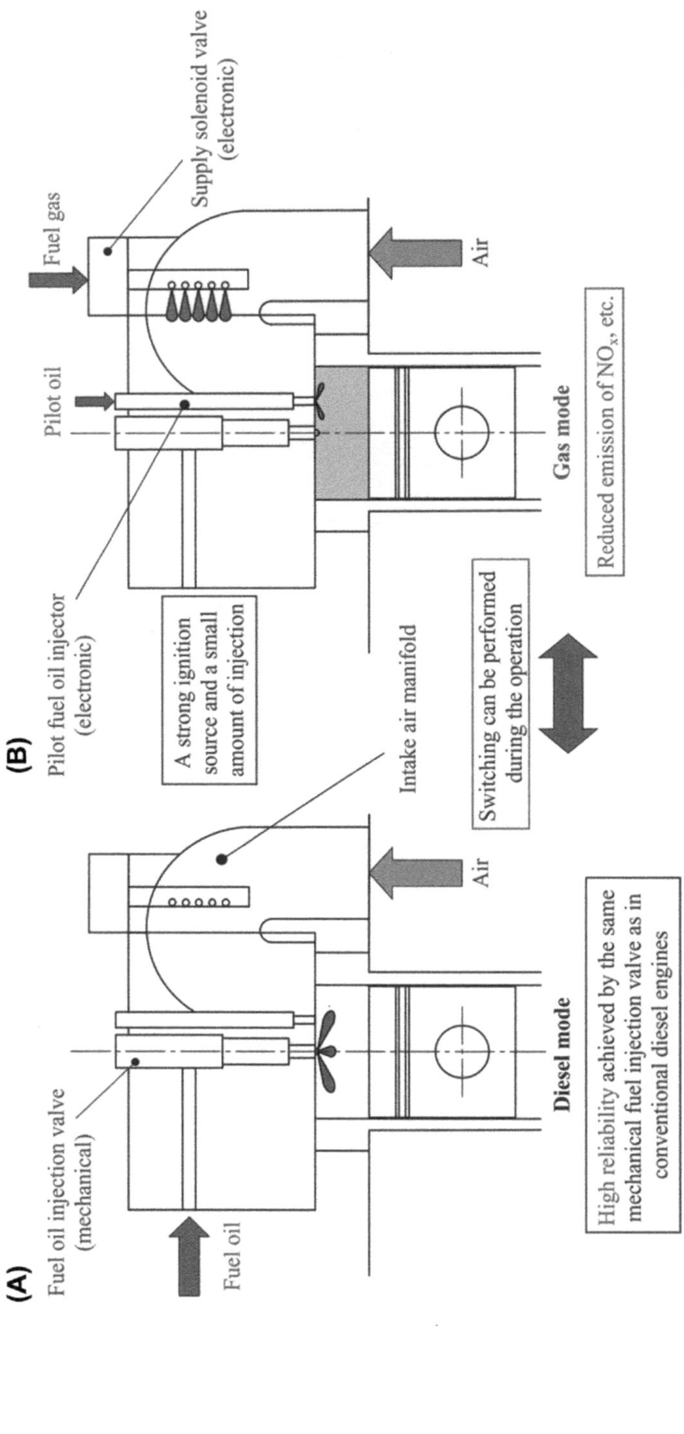

Figure 5.1.2 Schematic of diesel and gas modes in the dual-fuel engine [3]. (A) Diesel mode and (B) gas mode.

The engine operational performance tests of the marine dual-fuel engine (6MG28AHX-DF) [3] at different atmospheric temperatures confirm that the transient performance is improved by the aforementioned antiknocking techniques, which is performed by adjusting the common rail injection timing properly if knocking occurred. In the test, when the output power of the engine is increased from the idle speed to the rated speed, transient performance of the engine rotating speed to the rated speed is investigated. At a relatively higher atmospheric temperature of 37°C, it is achieved by adequate air intake. In addition, the engine should adapt to rapid load changes in the operation pattern; the control system should be optimized for the instant switching of engine operations from gas to diesel modes. The tests demonstrate that the operational performance in the gas mode agreed with that in the diesel mode, i.e., the engine can adapt to rapid load changes and switching during engine operations. Therefore, the engine operation in the gas mode can be applied to ship propulsion systems just like conventional diesel engines. On NO_x in the exhaust gas, the dual-fuel engine satisfied the IMO Tier II NO_x emission regulation in the diesel mode and Tier III one in the gas mode. In addition, the operation in the gas mode realized a CO_2 reduction of 19% in comparison with that in the diesel mode. The results concluded that the dual-fuel engine can be used successfully for low-emission ships.

5.1.2 NO_x emission control using plasma chemical hybrid process

In the authors' laboratory at the Osaka Prefecture University, the nonthermal plasma (NTP) environmental application technology has been attempted to develop, which is new environmental protection since 1998. The primary research topics for exhaust gas cleaning are as follows: NO_x control for industrial boiler emission and NO_x control for glass melting furnace emission. In the following, these two technologies are explained, which can be applicable for future marine diesel engine NO_x emission control.

NO_x control for industrial boiler emission

The number of small boilers using city natural gas, heavy oil, and waste oils has been increasing annually, and more stringent regulations for NO_x emission are anticipated for the reduction in environmental NO_x concentration. A suitable flue gas treatment system for small boilers will be required soon. Several studies have been conducted on laboratory-scale NTP-chemical hybrid processes for the removal of NO_x from gases emitted from various stationary sources. The authors propose a plasma chemical hybrid process (PCHP) consisting of an indirect NTP treatment and a wet chemical one [4]. The NTP treatment is performed according to the reaction (5.1) and the wet chemical treatment is performed according to the reaction (5.2).

$$NO + O_3 \rightarrow NO_2 + O_2 \tag{5.1}$$

$$2NO_2 + 4Na_2SO_3 \rightarrow N_2 + 4Na_2SO_4 \tag{5.2}$$

Kim et al. [5] reported a new PCHP using a sodium sulfide (Na_2S) wet chemical scrubber. High performance NO_x removal in flue gas from semiconductor industries was successfully performed. In the present NO_x treatment from boiler, nontoxic Na_2SO_3 solution was used.

Fig. 5.1.3 shows an overview of the plant for the application of PCHP to a boiler. The flue tube industrial stationary boiler (Takao Iron Works Co., Ltd., Japan) has an original rotary burner for gas and/or oil fuel. It can be operated using either city natural gas (13A) at 157 Nm^3/h, where N means the standard state (0°C and 101.325 kPa) or the A heavy oil at 171 L/h and waste oils. The boiler has a steam generation rate of 2.5 ton/h. Two sets of silent discharge-type ozonizers (EW-90Z, Ebara Jitsugyo Co., Ltd., Japan) equipped with a pressure swing adsorption (PSA) oxygen generator are employed to generate O_3.

Fig. 5.1.4 shows the structure of the electrical discharge section of the plasma reactor in the ozonizer. A coaxial NTP reactor or silent discharge-type reactor is used. O_3 gas is generated after O_2 gas by PSA passes through the 30 plasma reactors inside the ozonizer and then injected into a flue gas

Figure 5.1.3 Plasma hybrid clean boiler system.

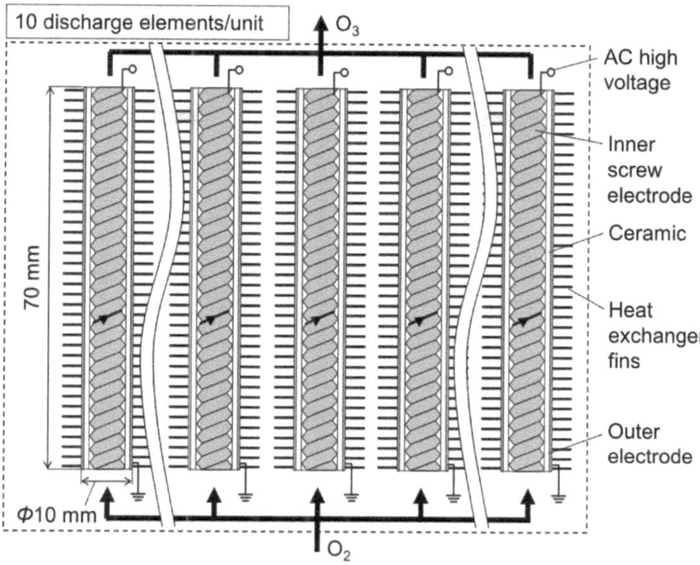

Figure 5.1.4 Structure of the discharge section in the ozonizer.

duct for NO oxidation. The O_3 gas flow rate is constant, i.e., 0.9 Nm^3/h. When the discharge power is 1.5 kW and the PSA power is 1.6 kW, 90 g/h of O_3 is generated at the maximum with a concentration of 4.7%, with an energy efficiency of 29 g/kWh. High-concentration O_3 gas is injected into the duct and diluted to oxidize almost all the NO to NO_2 according to reaction (5.1). The flue gas is then introduced into the scrubber to reduce NO_2 to N_2 according to reaction (5.2). The experimental results and discussion are shown next. To confirm the NO_x removal performance, the mixed oil of waste vegetable oil and heavy oil and the city natural gas are tested as the boiler fuels.

Fig. 5.1.5 shows the time-dependent NO and NO_x concentrations at MP1 (measurement point one at the exit of the boiler) and MP3 (measurement point one at the exit of the exhaust gas cleaning). The NO_x levels are measured by NO_x analyzers. The experiment is performed with a daily start and stop operation. The average flow rate of the city natural gas is 52 Nm^3/h and that of flue gas is 718 Nm^3/h at an O_2 concentration of 3.3–4.0%. The average rates of O_3 injection and sodium sulfite (Na_2SO_3) supply are 86 g/h and 63 mol/h, respectively. NO_x concentrations at MP1 are within 45–50 ppm. This small difference in NO_x daily is due to the difference in the O_2 concentrations at MP1 between 3.3% and 4.0%. The NO_x concentrations at MP3 are within 5–8 ppm, which are equivalent to a NO_x removal efficiency of more than 85% over an operating time of 23 h. The NO_x removal performance achieved is satisfactory and stable.

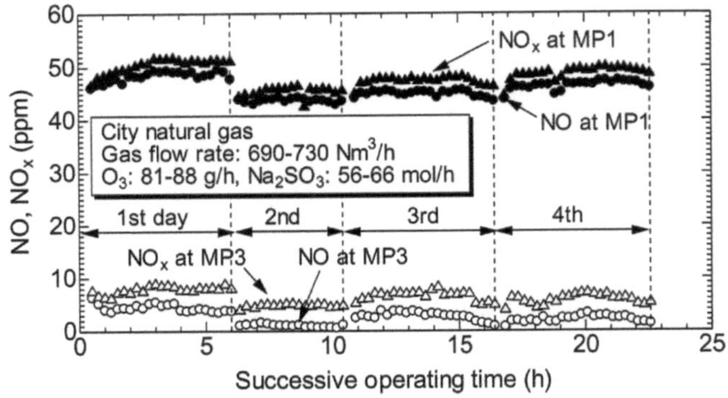

Figure 5.1.5 Time-dependent NO and NO_x concentrations at MP1 (boiler exit) and MP3 (exit of aftertreatment) during a four successive day operation in firing city natural gas.

The specific gravity of the scrubbing solution increases gradually and is stable at approximately 1.08, which is an allowable value as a control index.

NO_x control for glass melting furnace emission

For the exhaust gas generated in a glass manufacturing system or glass melting furnace because it is difficult to apply an SCR catalyst for NO_x reduction owing to the corrosive impurities contained in the exhaust gas, there have been almost no effective decomposition of nitrogen oxide (De-NO_x) methods applicable. Currently, NO_x is reduced using lower air-ratio combustion at the furnace. Therefore, research has been conducting aimed at applying the PCHP to this problem. In this subsection, the current status of the exhaust gas treatment of a glass melting furnace using PCHP, the principle of PCHP for the glass melting furnace, and the test results obtained in actual manufacturing apparatus are introduced. This cleaning technology can be applied in the future for the treatment of marine diesel engine exhaust gas including NO_x, SO_x, and PM.

Fig. 5.1.6 shows a schematic of the glass bottle manufacturing system and the dry exhaust gas treatment process [6]. Raw materials (silica sand, limestone, soda ash, sodium sulfate [Na_2SO_4], etc.) are melted in a glass melting furnace at approximately 1500°C using fossil fuel of C heavy oil and city gas. In a bottle manufacturing system, molten glass is formed into a bottle shape using a mold, gradually cooled, and generated into a glass bottle product after inspection. However, because exhaust gas discharged through the combustion process contains air pollutants such as SO_x derived from the raw materials and fuel, NO_x generated from high-temperature air combustion, and dust of scattering materials, the following exhaust gas treatment facilities are installed to reduce the environmental load of such substances.

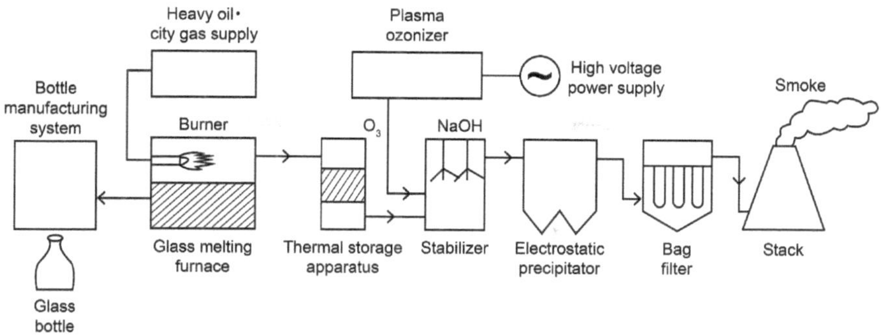

Figure 5.1.6 Schematic of glass bottle manufacturing system and dry exhaust gas treatment process.

The exhaust gas from a glass melting furnace passes once through a thermal storage apparatus and is introduced into the reaction tower, called stabilizer. In the stabilizer, an aqueous solution of sodium hydroxide (NaOH) is sprayed, and SO_x in the exhaust gas is removed. The removed SO_x changes into Na_2SO_3 and is recovered as Na_2SO_4 through oxidation. After the process, it is used again as a raw glass material. The fine particles of Na_2SO_4 and dust particles generated are removed using an electrostatic precipitator and a bag filter located downstream. The cleaned exhaust gas is then discharged from the stack.

However, because the exhaust gas contains a large amount of adhesive dust derived from the raw materials contained in it, as well as a high concentration of SO_x, the introduction of NO_x removal equipment has not progressed because SCR is difficult to be used. The regulated value (450 ppm in Japan) is realized using a lower air ratio combustion technology. However, because NO_x reduction through lower air ratio combustion increases the fuel consumption, a technology for reducing NO_x with an energy-saving capability has been desired. Therefore, PCHP technology for a glass melting furnace for De-NO_x has been developed. As shown in Fig. 5.1.6, ozone gas generated by the plasma ozonizer, shown in the figure, is injected into the stabilizer while cooling with the solution can be achieved.

The De-NO_x is based on reactions (5.1) and (5.2). However, Na_2SO_3 as a reducing agent for NO_x is obtained through the following reaction (5.3) of NaOH and SO_2 in the exhaust gas.

$$SO_2 + 2NaOH \rightarrow Na_2SO_3 + H_2O \tag{5.3}$$

Two types of discharge plasma ozonizers were used at the same time in carrying out the pilot-scale demonstration test. One is a silent discharge type of ozonizer, which is equipped with a pressure swing adsorption (PSA) type oxygen generator as an oxygen supply source, and is called ozonizer A. The other is a new surface discharge type of ozonizer, which uses oxygen cylinder curd or liquid oxygen as the oxygen supply source, and is called ozonizer B. Ozonizer A (EW-90Z, Ebara Jitsugyo Co., Japan) uses air-cooled silent discharge plasma and is equipped with a PSA O_2 generator. Ozonizer B is driven by six sets of high-voltage power supplies and is supplied with O_2 gas from gas cylinders. Twelve NTP reactors (HCII-OC 70 × 12, Masuda Research Co., Ltd.) are equipped for a single NTP apparatus. In this demonstration test, three ozonizers A and four ozonizers B were used simultaneously. The maximum amount of generated ozone is 1.22 kg/h and 2.1% (= 44 g/Nm^3) of ozone concentration at a maximum power consumption of 24 kW.

Test results of the exhaust gas treatment of a glass manufacturing system are described. For the exhaust gas treatment applied to a glass melting furnace mainly used in developing countries, semidry-type desulfurization equipment, as shown in Fig. 5.1.6, is used more frequently than wet-type equipment, the reason for which is that wet treatment requires a final treatment of the wastewater, and the appearance of white smoke from steam discharged from the stack may cause certain problems; thus, the amount of aqueous chemical solution sprayed in a reaction tower should be reduced for dry treatment. For this reason, it is preferable to use dry-type equipment in which the exhaust gas is completely dried. A pilot-scale dry plant combining PCHP in a semidry-type desulfurization apparatus of a glass melting furnace was constructed, and the removal of NO_2 by carrying a neutralizing agent ($NaHCO_3$) in a bag filter downstream was conducted [6]. Simultaneous dry decomposition of sulfur oxide (De-SO_x) and De-NO_x test using an exhaust gas volume flow rate of approximately 12,700 Nm^3/h was also applied.

In this glass melting furnace, combustion is performed through the switching of the paired thermal storage apparatus and burners at fixed time intervals. In this case, Na_2SO_4 produced by reaction (5.3) in the stabilizer through the oxidation of Na_2SO_3 becomes PM and is collected by an electrostatic precipitator located downstream. This is necessary to maintain the temperature of the exhaust gas at the outlet of the reaction tower within the range of 200–250°C to release completely dry exhaust gas and prevent a reduction of the dust collection rate. Therefore, it is necessary to keep the amount of NaOH solution sprayed to a properly low amount. However, it is known that when the exhaust gas temperature exceeds 150°C, the performance of NO oxidation through O_3 injection deteriorates. For this reason, it is important to efficiently supply ozone to a local cooling region at less than 150°C. The ozone spray nozzle is directed toward the center of the cooling region using solution droplet vapor evaporation. As a result, it is confirmed that the outlet temperature of the reaction tower can be maintained at 200°C or higher while creating a local cooling region temperature of about 70°C.

The overall results of the performance tests are presented in Table 5.1.1. In the five tests named T1 to T5, various changes in the conditions occur. The maximum removal efficiencies of NO, NO_x, and SO_2 are 28.1%, 4.4%, and 64.1%, respectively. In each case, NO decreases from 160 to 115 ppm, NO_x decreases from 158 to 151 ppm, and SO_2 decreases from 142 to 51 ppm. Because the test was conducted on a pilot scale, the amount of ozone injected was not enough, and the reduction efficiency was therefore not very high. An optimization test of the reaction conditions for reducing NO_x is progressing. However, the removal efficiency of NO_x in the bag filter using $NaHCO_3$ was 45% at maximum. As a result, it was confirmed that the dry De-SO_x and De-NO_x treatment of the exhaust gas of a glass melting furnace using PCHP is possible. In the future, a practical full-scale application and the research and development for a cleaning apparatus of marine diesel engine exhaust gas will be prepared.

5.1.3 Microwave plasma reduction of marine diesel NO_x and particulate matter

A reduction of NO_x and PM of the exhaust gas of a marine diesel engine using microwave NTP has been studied [7,8]. Microwave plasma generation can remove the NO in a typical two-stroke marine diesel engine. Fig. 5.1.7 shows a schematic of a pilot-scale NTP reactor using two microwave generators (No. 1). In the reactor, microwave energy is launched into the multimode cavity (No. 4) through the stub tuners (No. 2) and the waveguides (No. 3). In the multimode cavity, passive electrodes are installed to generate microwave plasma. Exhaust gas from an engine passes through the multimode cavity via the gas inlet/outlet (No. 5). The passive electrodes, which consist of a needle electrode, saw tooth electrode, and cross saw-tooth electrodes, are used to generate a high electric field region within a microwave environment where high energetic electrons of 1–3 eV are produced for the generation of NTP. The generator-set diesel

Table 5.1.1 Test results of dry De-NO_x system.

Test number	Flow rate of exhaust gas m^3N/h	Averaged concentrations before/after treatment ppm[a]			Removal efficiency %			Injected ozone kg/h	Added $NaHCO_3$ kg/h
		NO	NO_x	SO_2	NO	NO_x	SO_2		
T1	12,700	160/115	160/153	135/67	28.1	4.4	50.4	1.2	0
T2	12,700	160/119	158/151	133/80	25.6	4.4	39.8	1.2	0
T3	12,700	164/150	162/166	141/73	8.5	0	48.2	0.0	0
T4	12,700	163/151	161/165	142/51	7.4	0	64.1	0.0	30
T5	12,700	164/134	162/156	140/52	18.3	3.7	62.9	1.0	30

[a]The average concentrations are converted into the concentrations at an oxygen concentration of 15%.

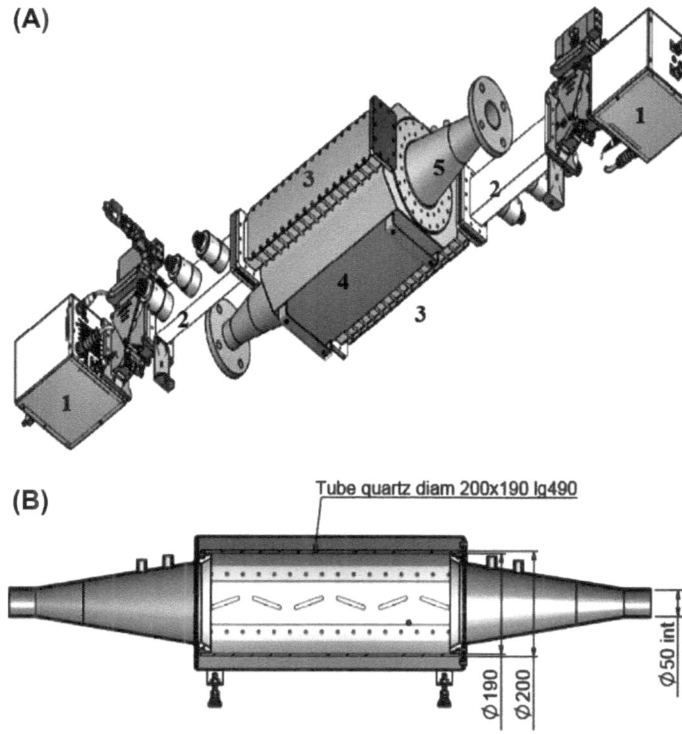

Figure 5.1.7 Schematics of pilot-scale nonthermal plasma reactor using two microwave generators [8]. (A) Three-dimensional schematic showing 1. microwave generators, 2. stub tuners, 3. waveguides, 4. multimode cavity, and 5. gas inlet/outlet. (B) Side view of the nonthermal plasma reactor, showing the reactor dimensions and slot arrangement.

exhaust gas is used to test the pilot-scale microwave plasma reactor. An experimental setup that was used in the research includes a pilot-scale NTP reactor using two microwave generators, 2-kW generator-set diesel generator, gas measurement systems, a high-voltage 30-kV supply, a gas flow rate meter, and a data-logging system. Two 2-kW microwave generators operating at 2.45 GHz are used to supply the required microwave energy into the microwave cavity through two slotted waveguides. The exhaust gas from a generator-set diesel engine contains a higher percentage of NO than NO_2, which is very similar to the exhaust from a typical two-cycle diesel marine engine.

The experimental results show that almost 100% removal of NO is possible for an exhaust gas flow rate of 60 L/s at an exhaust gas temperature of 40–70°C. They also show that microwaves can significantly remove soot particles of PM having diameters of 10–365 nm entrained in the exhaust gas of a 200-kW marine diesel engine with a 40% engine load and a gas flow rate of 130 L/s. Microwaves irradiations demonstrate a PM reduction of up to 50% without the generation of plasma by high temperature and up to a 90% reduction with the generation of plasma. Because it requires a fine tuning of the electrode position, the major challenge in these experiments is igniting the desired plasma and sustaining it using passive electrodes for a longer period, which is influenced by many factors including the gas flow rate, the geometry of the reactor, and the microwave power applied.

5.1.4 Direct nonthermal plasma reduction of marine diesel NO_x and particulate matter

A combined approach to reducing NO_x and PM emitted by medium-speed marine diesel engines using NTP reactors has been studied in Poland [9]. Fig. 5.1.8 shows experimental, analyzer, and acquisition system setup along with an NTP unit in an engine exhaust gas bypass. In the figure, marine diesel engine exhaust gas passes through a silencer via the inlet and outlet. A bypass channel is installed at the silencer. An NTP module is installed in a bypass channel (with about a 10% flow rate of that in the main duct) of the main exhaust system of a four-cycle marine diesel engine (6AL20/24, HCP Sulzer, Poland) for the main propulsion or generator sets of a small ship. The output power of the engine is 397 kW, and the rotating speed is 720 rpm. A test bench construction enables direct exhaust stream exposure using main exhaust bypass to a plasma reactor without any additional components such as coolers and orifices that could alter the exhaust gas properties. The design of the employed reactor relies on a dielectric barrier discharge consisting of 36 parallel electrode rods positioned to allow movement of a gas stream through the assembly. The plasma is driven by a novel power supply based on a series–parallel resonance circuit topology, which permits high-voltage and frequency operations. The gas component concentrations, such as CO, CO_2, NO, NO_2, N_2O, NO_x, O_2, soot, soluble organic fraction (SOF and adsorbed hydrocarbons), and total PM, are measured using gas analyzers. Throughout the measurement trials, the gas composition in the engine exhaust gas system is monitored using a gas analyzer before the gas passes through the NTP reactor, whereas two gas analyzers are fitted at the NTP reactor outlet. The data of gas component concentrations and engine monitoring data are sent to the data server. A set of thermocouples is used to estimate

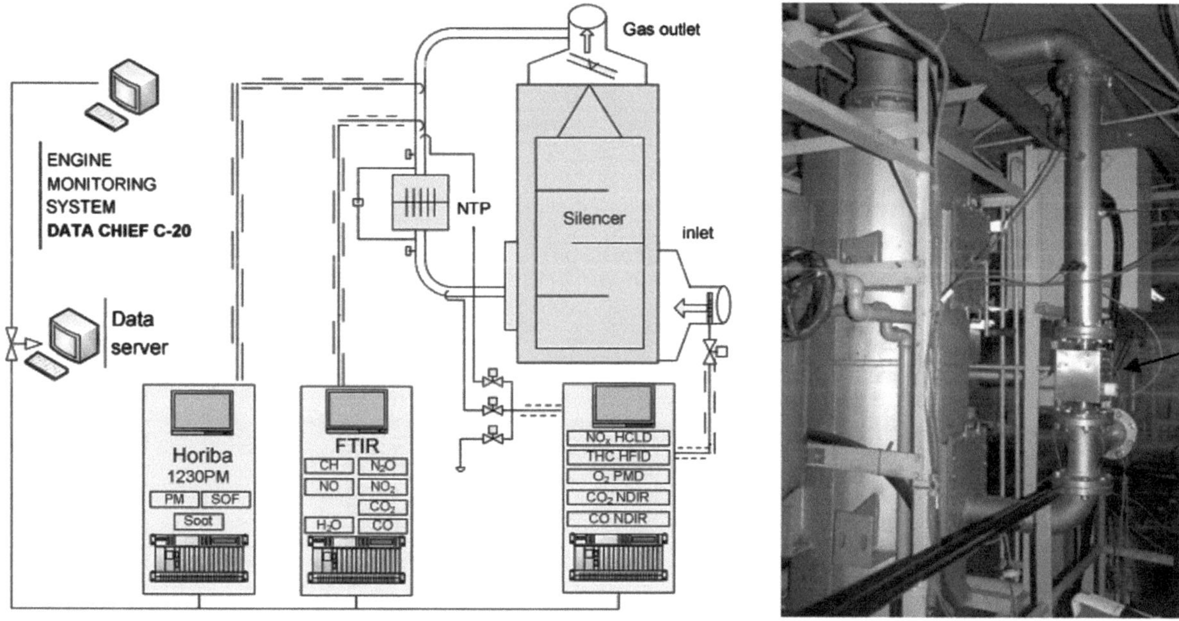

Figure 5.1.8 Experimental, analyzer, and acquisition system setup along with nonthermal plasma unit in engine exhaust gas bypass [9].

the temperature in the NTP module. The trial schedule allows the reactor to heat up and reach adequately stable conditions for the engine exhaust system and outlet gas conditions. The NTP reactor temperature cannot be adjusted because it depends on the exhaust gas temperature of the engine. During engine tests, this temperature varies from 150°C when idling to 350°C under a full engine load. This dependence is used to investigate the influence of the operating temperature on the related oxidation processes.

Results from the implementation of NTP reactor in the exhaust duct of a test bench diesel engine demonstrate considerable improvements in the gas composition and PM emissions at extremely low energy densities (up to 16 J/dm^3). Significant effects on the conversion of NO and PM under realistic engine exhaust conditions are noted. The removal rate of the unbalanced PM amount is found. The removal rate reaches a maximum value of about 12% under 265 W. The total PM removal efficiency is 1.7 g/kWh at an active input power of 180 W. Partial oxidation of NO to higher nitride oxides such as NO_2 is observed, as is the conversion of PM particles into gaseous SOF and soot. Partial oxidation of NO is suggested as conditioning for NO_x catalyst—based reduction processes at temperatures of below 200°C. These results suggest a two-stage process. Namely, the release of soluble organic fraction (SOF) by heat, followed by oxidation through the reactive oxygen species in the plasma, or through the direct decomposition of the polycyclic aromatic hydrocarbon (PAH) in the SOF particles.

5.2 Advanced SO_x control technologies

5.2.1 Classifications of SO_x control technologies

Three techniques exist for controlling SO_x from engines: (1) using low-sulfur fuel, (2) using aftertreatment or scrubber apparatus, and (3) using dual-fuel (LNG/oil) engines. Although some of these techniques are explained in the preceding texts in the following, from the practical view-point, each feature with possible near-future applications for SO_x control technologies is explained [10].

The low-sulfur fuel technique (1) has been realized in the past for oil-fueled marine diesel engines. Although it is the simplest and effective method, the following two problems exist. One is that the price difference between high-sulfur fuel and low-sulfur fuel is large. Therefore, the influence on operation cost is high. Another is that although the response to regulation is possible by switching between high-sulfur fuel and low-sulfur fuel at the time of entering and leaving the Emission Control Area (ECA), the flow rate must be switched gradually because the temperature difference among both fuels is large. For a high-sulfur fuel, the usage temperature is higher because heat is added to it to decrease the viscosity. Rapid changes may damage parts of the fuel system in the engine.

Using an aftertreatment apparatus (2) is a powerful way of removing SO_x, and it has already explained in the preceding Chapter 4. Reducing SO_x that is emitted from an engine is considered equivalent to using low-sulfur fuel. After it is approved by the regional government to install aftertreatment apparatus and desulfurize SO_x, the regulation can be satisfied while using inexpensive high-sulfur fuel. Two methods exist for the aftertreatment apparatus: the wet type using showering aqueous solution and the dry type using solid absorbents such as activated carbon. In the wet type, three primary types exist: one is a method of circulating clean water with/without adding chemicals for washing exhaust gas, other is a method of discharging water after treatment using seawater, and the other is a method capable of switching both functions. When the solution is drained, the regulatory values of pH, polycyclic aromatic hydrocarbon (PAH), turbidity, and concentrations of nitrate salts must be observed. On the other hand, although no wastewater is present in the dry method, the solid adsorbent pellets must be loaded and the used pellets must be disposed at the port. In both two methods, the aftertreatment apparatus is large and heavy; therefore, the mass balance of the installation space at the ship must be considered. Further, considering sailing time inside the ECA, the price difference between high- and low-sulfur fuels, the size of the scrubber, the installation cost, etc., the optimum design for each ship must be determined.

Using dual-fuel (oil/gas) engine (3) is another powerful way of controlling SO_x. SO_x emissions can be reduced to almost zero using a gas fuel that does not contain sulfur, and NO_x emission also greatly decreases. In recent years, with regard to natural gas, shale gas—taking technology has considerably advanced, and the development of gas fields is progressing especially in the United States. In addition, a dual-fuel engine using a fuel with a small environmental load and capable of switching from gas fuel combustion to a liquid fuel one has been developed by many engine manufacturers and is being used for ships. In Subsection 5.1.1, a duel fuel engine system is explained from the viewpoint of NO_x reduction. The features will be introduced based on the performance of another dual-fuel engine system compliant with environmental regulations from the viewpoint of SO_x reduction.

5.2.2 Using dual-fuel engine

By switching from liquid oil fuel to natural gas fuel, the dual-fuel engine has the following four advantages: (1) environmental compliance with a lower fuel cost: Within the ECA for NO_x, corresponding to exhaust gas regulation, the gas fuel mode of operation is effective with no exhaust gas aftertreatment apparatus required. The NO_x regulation can comply with the IMO Tier III regulation. On the SO_x regulation, both SO_2 and PM emissions are almost zero. However, outside the ECA, the engine meets the NO_x IMO Tier II regulation in diesel mode with low-cost oil fuel. (2) CO_2 reduction: In the gas-mode operation, CO_2 emissions are reduced compared to diesel engines with liquid oil fuel. Fig. 5.2.1 shows the volume-based emission comparison when MDO and gas fuel are used. CO_2 and NO_x emissions are reduced by approximately 23% and 90%, respectively, compared with diesel engines with liquid oil fuel. It is also confirmed that the SO_2 and particulate emissions are almost zero. (3) Fuel redundancy: The flexibility in selecting a fuel depending on price is possible. Furthermore, even if one fuel is depleting, the ship can be operated with the other fuel. (4) Size reduction of the LNG tank: It can be small compared with the gas fuel only ship.

Table 5.2.1 shows the main specifications of a dual-fuel marine diesel engine (6DE28DF, Daihatsu Diesel MFG Co., Ltd.). It is a four-cycle engine used for both ship propulsion and power generation inside ships.

The operation test results of the dual-fuel engine are explained. Two operation modes exist for the engine. In the diesel mode, it is compliant with the IMO NO_x Tier II regulation and cleared the NO_x regulation

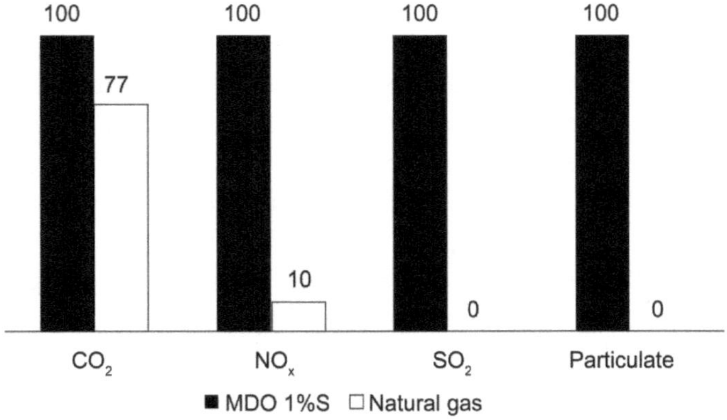

Figure 5.2.1 Emission comparison between marine diesel oil including 1% sulfur and natural gas fuel [10]. *MDO*, marine diesel oil.

Table 5.2.1 Main specifications of the marine diesel engine, 6DE28DF [10].

Bore	mm	280
Stroke	mm	390
Number of cylinders	—	6
Rotation rate	min^{-1}	720/750
Regulated power	kW	1730

value by the combustion improvement. In the gas mode, although SO_x is almost zero, considerable adjustment is necessary to cope with the IMO NO_x Tier III regulation because of thermal NO_x formation. The following items are the primary adjustment factors: air-to-fuel ratio, micropilot fuel injection timing and quantity, gas supply pressure, and intake valve closing timing. In the gas mode, a high efficiency is typically obtained in the air-to-fuel ratio of approximately 2.1—2.2 (lean-burn combustion with turbocharger). These factors are optimally adjusted in the test. Consequently, the IMO NO_x Tier III regulation value was cleared in the gas mode.

The switching features of the operation mode from gas to oil and oil to gas are very important. When a trip continues in the gas mode, it is sometimes necessary to instantaneously switch to the diesel-oil mode. It was confirmed that mode switching is possible by stopping the operation of the gas injection valve and instantly injecting the diesel governor output without changing the engine speed and load. An example of switching from gas to oil is shown in Fig. 5.2.2. The time-dependent rotational speed

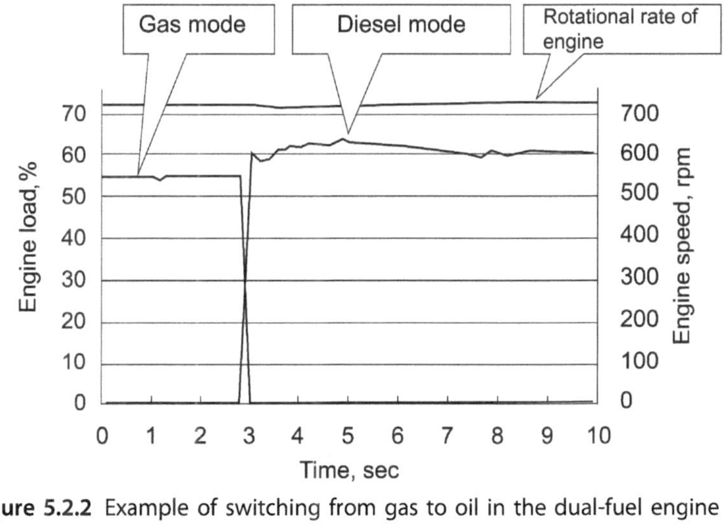

Figure 5.2.2 Example of switching from gas to oil in the dual-fuel engine [10].

and power of the engine are shown in the graph. As shown in the figure, switching is completed in only 0.2–0.3 s, and the rotation speed and power are also stable.

In the case of switching from the diesel mode to the gas mode, if the amount of gas supply rapidly increases, a fuel-rich state occurs to induce knocking. Therefore, instantaneous switching is difficult; however, it is confirmed that stable switching is possible by increasing the gas supply amount gradually while maintaining the air-to-fuel ratio at which knocking does not occur. In addition, because the stable operating region is wider on the low-output side, switching in a shorter time is possible.

5.2.3 Economic analysis of liquefied natural gas-fueled and oil-fueled ships

As described in the previous subsection regarding the dual-fuel engines, LNG-fueled engines have lower emissions of NO_x, SO_x, and PM than oil-fueled ones. However, LNG engines have higher capital cost and operation cost, owing to the price difference between LNG and oil. If NO_x emission aftertreatment of SCR is required, the urea solution cost should be considered. The market prices of fuels and urea in 2011–12 are summarized in Ref. [11] and shown here. The price of oil fuel depends on the sulfur content of fuels. The price of low-sulfur fuel oil (LSFO) or MDO (0.1% or 0.5% sulfur) is estimated as 1100 US$ or USD (US dollar)/ton. The price of LNG is 440 USD/m^3. The price of 40% urea solution (=800 USD/ton) is almost the same as that of heavy fuel oil (HFO) (=720 USD/ton). Because the import price per unit energy of LNG is approximately 20 USD/mmBTU (million British Thermal Unit, 1 BTU = 1.055 kJ) and that of heavy oil is approximately 17 USD/mmBTU in 2012, the price of LNG fuel is slightly higher than that of the oil fuel. From these reasons, economic analysis of international transocean merchant ship's sailing considering capital and operation costs of the LNG and oil-fueled engines is interesting because oil-fueled engines require urea for NO_x reduction using SCR, although the price of oil fuel is lower than LNG fuel.

The economic analysis of transocean LNG-fueled container ships, sailing between Asia and Europe, was performed by Adachi et al. [11], and the results are explained in this subsection. From the viewpoint of the transportation economy, a comparison was conducted with the existing container ship fueled by bunker oil. Bunker fuel or bunker crude is technically a type of fuel oil used aboard ships; it corresponds to heavy oil in this case. For the targeted ship specification, an existing 9300 TEU oil-fueled container ship operating between Japan and Europe via the Suez Canal is treated. The specifications of the container ship are as follows: maximum load capacity = 9300 TEU, actual load capacity = 9144 TEU, length L = 332.15 m, length between perpendiculars L_{pp} = 315.0 m, width W = 45.2 m, depth D = 26.8 m, deadweight (DW) (planned) = 71,200 ton, DW (summer) = 88,200 ton, maximum output of main engine = 58,400 kW, total output of four auxiliary engines = 2900 kW, and low-speed direct-coupled propulsion system. The operation of the oil-fueled ship is as follows: days for a round trip = 70 days, mean speed = 19 knot, engine output = 22,000 kW, and engine load = 38%.

Based on the specifications and the operation, the following virtual future projects A–C of the container ships, which complies with the IMO Tier III emission regulation, were designed as follows. Here, the ECA is assumed as the entire Mediterranean Sea.

1. Project A: An oil-fueled low-speed diesel direct-coupled propulsion (LD) ship with an SCR device is examined. The specifications are as follows. Main engines and propulsion: engine speed = low, engine type = diesel, propulsion system = direct, number = 1 unit, total output = 58,400 kW, propulsion power = 22,000 kW, and fuel rate = 174 g/kWh. Auxiliary engine: engine speed = medium, engine type = diesel, number = 4 units, total output = 2900 kW, and fuel rate = 217 g/kWh. Fuel: In the ECA, LSFO (0.1% sulfur) is used; off the ECA, LSFO (0.5% sulfur) is used. For the boiler, LSFO is used. The SCR for the main engine is installed, but not for the auxiliary engine.
2. Project B: An LNG-fueled LD ship is examined. Dual-fuel low-speed diesel engine is installed as the main engine, LNG is the main fuel, and LSFO is the pilot fuel. The specifications are as follows. Main engines and propulsion: engine speed = low, engine type = DF, propulsion system = direct, number = 1 unit, total output = 58,400 kW, propulsion power = 23,100 kW, fuel rate = 174 g/kWh. Auxiliary engine: engine speed = medium, engine type = DF gas, number = 4 units, total output = 2700 kW, fuel rate = 217 g/kWh. Fuel: In the ECA, LNG and LSFO (0.1% sulfur) (10% of total fuel) are used. For the boiler, LNG is used. The SCR for the main engine is installed, but for the auxiliary engine, it is not installed.

3. Project C: An LNG-fueled medium-speed diesel-electric propulsion (ME) ship is examined. LNG is utilized all the way around. The specifications are as follows. Main engines and propulsion: engine speed = medium, engine type = DF gas, propulsion system = electrical, number = 4 units, total output = 58,400 kW, propulsion power = 23,100 kW, and fuel rate = 185 g/kWh. Auxiliary engine: engine speed = medium, engine type = DF gas, number = 1 unit, total output = 2700 kW, fuel rate = 193 g/kWh. Fuel: In both the ECA and off the ECA, LNG and LSFO (3% of total fuel) are used. For the boiler, LNG is used. The SCR for the main engine and the auxiliary engine is not installed.

An oil-fueled LD ship is additionally examined as a reference ship as Project Z. The specifications are as follows. Main engines and propulsion: engine speed = low, engine type = diesel, propulsion system = direct, number = 1 unit, total output = 58,400 kW, propulsion power = 22,000 kW, fuel rate = 170 g/kWh. Auxiliary engine: engine speed = medium, engine type = diesel, number = 4 units, total output = 2900 kW, fuel rate = 217 g/kWh. Fuel: Both in the ECA and off the ECA, HFO is used. For the boiler, HFO is also used. The SCR for the main engine and that for the auxiliary engine is not installed.

Based on the cost evaluation, the financial effects of projects A, B, C, and Z were analyzed. In the result, the net present value (NPV) curves are indicated in Fig. 5.2.3. It is noted that Project Z use the reference ship of the HFO oil-fueled without aftertreatment and cannot be compared directly with Project A-C. Therefore, the result does not appear in the figure. NPV is a measurement of profit calculated by subtracting the present values of cash outflows including an initial cost from the present values of cash inflows over a period of time. The horizontal axis expresses the time since the operation starts and the unit is year. The vertical axis expresses the NPV. The results of projects A, B, and C are shown in the figure.

Although the initial capital costs (absolute values of NPV at $t = 0$) for LNG LD and ME ships are higher, LNG ships are not only environmentally friendly but also economical from the viewpoint of the operational cost. The payback period is 7.9 years at an NPV of zero for project A, 7.4 years for project B, and 9.2 years for project C. After approximately 6 years, the initial cost gap between projects A and B are matched. After approximately 13 years, the initial cost gap between projects B and C is compensated. These results indicate that the LNG main engine is promising for the future driving apparatus. The result greatly depends on the prices of LNG and oil fuels. The calculations were carried out based on the 2011–12 market prices.

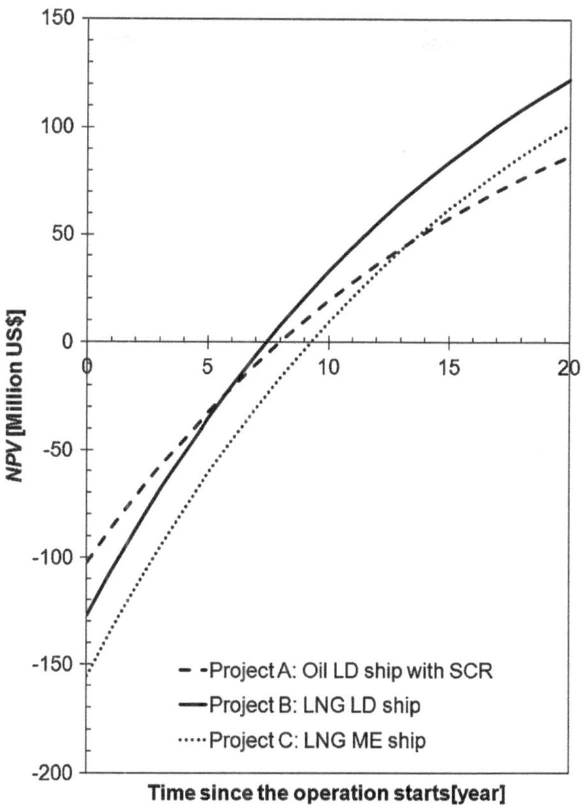

Figure 5.2.3 Time-dependent net present value (NPV) plots of projects [11]. *LNG*, liquefied natural gas; *SCR*, selective catalytic reduction.

5.3 Volatile organic compound emission controls

5.3.1 Volatile organic compounds

Volatile organic compounds (VOCs) mean any compound of carbon which participate in atmospheric photochemical reactions, excluding carbon monoxide, carbon dioxide, carbonic acid, metallic carbides or carbonates, and ammonium carbonate [12]. Moreover, VOCs are a generic term for organic chemicals whose boiling point is low and volatilize easily into the atmosphere at room temperature. In more detail, the World Health Organization (WHO) classifies such organic chemicals into four groups of very volatile organic compounds (VVOCs) volatile organic compounds (VOCs), semi-volatile inorganic compounds (SVOCs), and particulate organic matter (POM), as given in Table 5.3.1. VOCs include organic chemicals that negatively influence the environment and human health and

Table 5.3.1 Classification of organic chemicals as defined by World Health Organization [12].

Group of organic chemicals	Boiling point, T_b	Organic chemicals
Very volatile organic compounds (VVOCs)	$T_b < 50–100°C$	Methane (CH_4), formaldehyde (HCHO), acetaldehyde (CH_3CHO), dichloromethane (CH_2Cl_2), etc.
Volatile organic compounds (VOCs)	$50–100°C \leq T_b < 240–260°C$	Ethanol (C_2H_5OH), benzene (C_6H_6), toluene (C_7H_8), xylene ($C_6H_4(CH_3)_2$), etc.
Semi volatile organic compounds (SVOCs)	$240–260°C \leq T_b < 380–400°C$	Chlorpyrifos ($C_9H_{11}Cl_3NO_3PS$), dibutyl phthalate ($C_{16}H_{22}O_4$), bis(2-ethylhexyl) phthalate ($C_{24}H_{38}O_4$), etc.
Particulate organic matter (POM)	$380°C \leq T_b$	Benzo[a]pyrene ($C_{20}H_{12}$), polychlorinated biphenyl (PCB) ($C_{12}H_{(10-n)}Cl_n$), etc.

become, e.g., malodorous substances, causative agents of photochemical smog, and carcinogens.

5.3.2 Volatile organic compound emissions from ship

As for the internal combustion engines, vaporized VOCs from gasoline, that is, fuel for automobiles, are considered as an environmental issue [13]. VOCs also exist as unburnt fuel in the exhaust gas. In addition, while a huge amount of volatile products such as crude oil, gasoline, naphtha, distillate oil/kerosene, and the residual oil is transported by tankers, vaporized VOCs are released into the atmosphere from cargo tanks while loading, unloading, and transporting of the volatile products. Therefore, the causes of VOC emission are classified into two processes: (1) the emission from a marine diesel engine in the combustion process, and (2) the leakage and/or release of volatile products. VOC emission amounts depend on the physical and chemical characteristics of the volatile products and the ship size. VOC emissions are also influenced by the method of loading of the volatile products. For instance, during the loading operation of a volatile product, splash loading, wherein the filling pipe does not extend deeply in a tank of the volatile product, causes high levels of vapor generation and loss. On the other hand, submerged loading, wherein the filling pipe extends to the bottom of the tank of the volatile product, can reduce vapor generation. Emissions of VOC vapor from the tank occur in major types of vaporization losses: loading, ballasting, and breathing losses. Loading loss is caused by the emission of vaporized VOCs pushed by rising the liquid level of fuel when liquid products such as oil are loaded into the tank. As a result, vaporized VOCs in the tank are released into the atmosphere. Ballasting loss occurs in ships that do not have segregated ballast tanks, when ballast water is pumped into the tank. Transit losses and breathing losses are associated with the storage of a volatile product. Breathing loss is caused by the emission of vaporized VOCs released from a compartment through vapor contraction and expansion, owing to changes in temperature and barometric pressure. Some ships are equipped with control systems for these losses.

5.3.3 Volatile organic compound emissions regulations

VOC emissions are controlled in specific ports and terminals by a worldwide regulation. The regulation includes two objectives: (1) VOC emission control by a requirement to utilize a vapor emission control system (VECS) and (2) all tankers carrying crude oil have an approved and implemented ship-specific VOC management plan. The guidelines are provided by MEPC (Marine Environment Protection Committee). The related technical information on the systems and operations are provided also by MEPC. The vapor-collecting system shall comply with the IMO guideline named Standards for vapor emission control systems. Parties regulating tankers for VOC emissions shall submit a notification to IMO with information of the target tankers such as size and type. Regulation 15 in MARPOL convention Annex VI as shown in A.2. of Appendix states two aspects of VOC control [14]: In the first aspect, the control on VOC emitted to the atmosphere at particular ports or terminals is required to utilize a VECS. Both the shipboard and port are to be in accordance with the guideline of Standards for vapor emission control systems. In the second aspect, all tankers carrying crude oil are required to have an approved and effectively implemented ship-specific VOC management plan, addressing at least the points given in the regulation. This regulation not only applies to tankers but also applies to gas carriers only if the types of loading and containment system allow the safe retention of nonmethane VOCs on board or their safe return ashore.

5.3.4 Volatile organic compound emission control technologies

The actual aftertreatment of VOC emissions from ships should be considered. For instance, VOCs in a void tank, where fuel is installed, are recovered through a vapor-recovery pipe. The recovered VOCs are great pollutant source and should be treated. VOC recovery and VOC decomposition devices are explained in this subsection. Some of the technologies have not yet been applied in the maritime field. However, they are applicable and will be applied to satisfy the future stringent regulations.

Volatile organic compound recovery devices

VOC recovery devices usually use one of adsorption, absorption, and cooling–condensing methods. These technologies can also be combined in a system. In the adsorption method, VOCs are adsorbed by adsorbents with a high concentration when gas including VOCs passes through the adsorbents. As adsorbents, activated carbon, silica gel, alumina, and zeolite are used. These adsorbents are replaced or regenerated with desorption in actual uses. The adsorption–desorption processes are realized using the temperature swing adsorption (TSA) and pressure swing adsorption (PSA). TSA utilizes the balance shift between adsorption and desorption caused by temperature swing. PSA utilizes the balance shift between adsorption and desorption caused by pressure swing. The adsorbents are installed in the device as a fixed bed, fluidized bed, or honeycomb structures. In the absorption method, VOCs are absorbed in solvent or solution. In this method, a recovery process is additionally required. However, VOC separation from the solvent is not always easy. In a cooling–condensing method, gaseous VOCs are condensed into liquid under a dewpoint temperature by cooling and the liquid VOCs are recovered. Although this method is effective for cases where the VOC concentration is high, much energy is required for cooling the liquid to change phase.

Volatile organic compound decomposition devices

VOC decomposition devices use one of the combustion treatments, membrane separation, biological treatment, photocatalyst decomposition, NTP decomposition, etc. These methods can also be combined in a system. The combustion treatments, the membrane separation, and the NTP decomposition are explained in the next.

The combustion treatments include direct combustion, catalytic combustion, and heat storage combustion methods [15]. In a direct combustion method using fossil fuel, VOCs are burned and decomposed rapidly into CO_2, H_2O, etc., at a high temperature of 650–760°C. Although it is applicable for most VOCs, it is effective when the VOC concentration is high, more than 10,000 ppm of toluene equivalent concentration. When the VOC concentration is not high, a pretreatment for condensation is necessary to improve the decomposition efficiency. In catalytic combustion,

VOCs can be oxidized and decomposed at a relatively low temperature of 300–400°C by catalysts such as platinum (Pt) and palladium (Pd). However, catalyst poisoning, which is caused by chemical compounds of chlorine, sulfur, phosphorus, silicon, and so on, could reduce the oxidation and decomposition efficiency. It is effective for the case where the VOC concentration is high, more than 3500 ppm of toluene equivalent concentration. In the heat storage combustion method, VOCs can be oxidized and decomposed at a temperature of 800–900°C via heat storage materials such as ceramics. It is effective when the VOC concentration is high, more than 3500 ppm of toluene equivalent concentration.

In the membrane separation method, VOCs are separated from gas by a membrane. There are two types of membrane methods. One uses a porous membrane with molecular size holes, and another uses a nonporous membrane. In the porous membrane method, VOCs are physically separated selectively when the molecules of VOCs pass through the holes by the pressure difference on the membrane, that is, separation through the holes by the size of molecules. In the nonporous membrane method, molecules of VOCs are chemically dissolved selectively into the membrane and diffused there. Next, the molecules are released from the membrane into the other lower pressure side of the membrane.

In NTP methods, VOCs are decomposed into CO, CO_2, and H_2O in general by NTP. Combined methods of adsorption and NTP processes also exist. In this method, the VOCs in gas are firstly adsorbed in the adsorbents. Secondly, the adsorbed VOCs are released at a high concentration to improve the decomposition efficiency in the following NTP process. The VOCs at a high concentration are supplied into the NTP unit, and the VOCs are decomposed into CO, CO_2, H_2O, etc. by NTP.

5.4 $PM_{2.5}$ and health effects

5.4.1 Current status of understanding health effects

Although $PM_{2.5}$ and ozone are reported to cause health hazards including problems in the respiratory system, relevant knowledge is lacking. For example, Fig. 5.4.1 shows the atmospheric concentration of $PM_{2.5}$ in Japan [16]. In the figure, general measurement station means the one-year-averaged data measured at the general measurement stations, and automobile measurement station means the one-year-averaged data measured at the measurement stations specified for automobiles. To achieve the environmental standards for $PM_{2.5}$ in the atmosphere, the concentration of $PM_{2.5}$

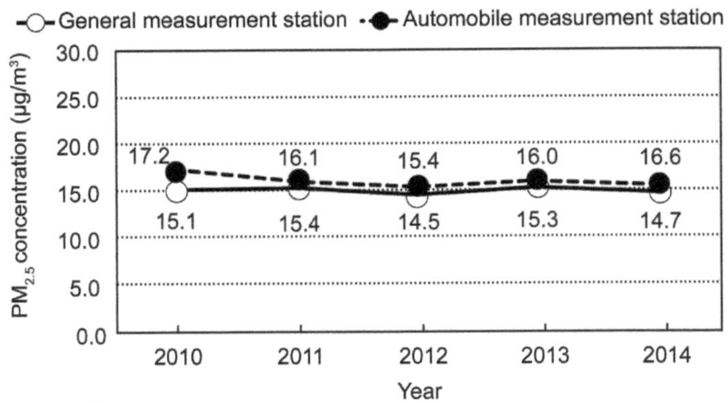

Figure 5.4.1 Concentration of $PM_{2.5}$ in Japan [16].

gradually decreases and is lower than that in neighboring Asian countries. According to some reports, a high value of $PM_{2.5}$, 80 $\mu g/m^3$, was recorded in South Korea in winter to spring because of transboundary pollution. $PM_{2.5}$ has a variety of sources such as secondary generation in the atmosphere and transboundary advection; therefore, the sources must be clarified to evaluate the health effects.

Table 5.4.1 presents the international comparison of $PM_{2.5}$ regulations. The standards of Japan and the United States of America are almost equivalent, 35 $\mu g/m^3$ in daily average although WHO standards is slightly stringent, 25 $\mu g/m^3$. The values of the annual average for $PM_{2.5}$ are lower, 10–15 $\mu g/m^3$. Furthermore, in Japan, a temporal guideline for alerting to atmospheric $PM_{2.5}$ concentration was discussed and determined at an expert meeting by Ministry of the Environment named Expert meeting on fine particulate matter ($PM_{2.5}$), February 2013. In the guideline, $PM_{2.5}$ concentration of level II alerting corresponds to the daily average value

Table 5.4.1 International comparison of $PM_{2.5}$ regulations [17].

		Japan	USA	World Health Organization
$PM_{2.5}$	Daily average	[a]35 $\mu g/m^3$	35 $\mu g/m^3$	25 $\mu g/m^3$
	Annual average	[a]15 $\mu g/m^3$	[c]12 $\mu g/m^3$	10 $\mu g/m^3$
PM_{10}	Daily average	[b]100 $\mu g/m^3$	150 $\mu g/m^3$	50 $\mu g/m^3$
	Annual average	—	—	20 $\mu g/m^3$

[a]Environment standards of $PM_{2.5}$ in Japan were announced in September 2009.
[b]Standard relating to suspended particulate matter (corresponding to approximately PM_7).
[c]Environment standards (annual average) of $PM_{2.5}$ in the United States of America revised on December 2012

more than 70 μg/m^3. On the other hand, $PM_{2.5}$ concentration of level I alerting corresponds to the daily average value equal or less than 70 μg/m^3. Because the daily average value of the environmental standard is 35 μg/m^3 as shown in the table, 70 μg/m^3 is larger than that. For level II alerting, an action guideline is as follows: Reduce unnecessary outings and long-term intense exercise outdoors as much as possible, and highly sensitive individuals are advised to act more carefully according to their physical conditions. Here, highly sensitive individuals mean people suffering from respiratory or cardiovascular diseases, children, elderly people, etc. For level I alerting, action guideline is as follows: Although behavior restriction is not necessary, the highly sensitive individuals may have an effect on health; changes in physical condition should be cared.

Fig. 5.4.2 shows the change in the concentration of photochemical oxidants including ozone, aldehyde, and peroxyacetyl nitrate (PAN). The meanings of the general measurement station and automobile measurement station are the same as those in Fig. 5.4.1. The data are daily-averaged one-year maximum values. The environmental regulation value for photochemical oxidants is 60 ppb (=0.06 ppm) in Japan. Although the concentrations are satisfied with the regulation, health effects to human respiratory system may occur by the sudden concentration increase by several 10 ppb. The concentrations of photochemical oxidants increase gradually. A weather condition has an effect on PM concentration increase near the ground surface where pollutants are likely to reach and stay on the ground owing to the temperature inversion layer. It is noted that the temperature inversion layer means an atmospheric layer where the temperature decrease with an increase in altitude is much less than that on the

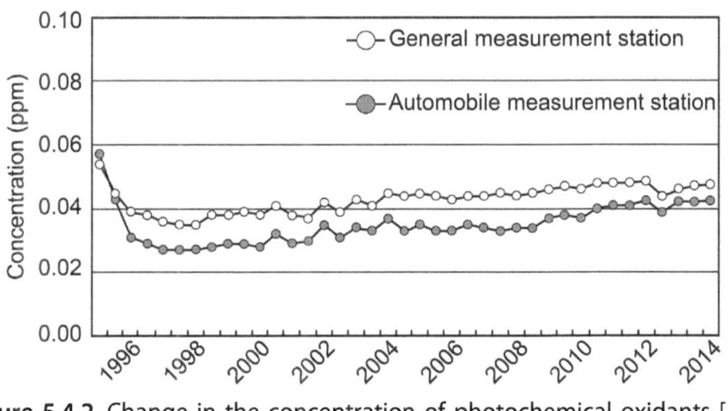

Figure 5.4.2 Change in the concentration of photochemical oxidants [16].

normal condition, and sometimes the temperature gradient becomes positive or the temperature increases with an increase in altitude. When such layer exits near the surface, the diffusion of pollutants toward the sky is suppressed and the concentration near the ground surface increases.

To reduce PM having aerodynamic diameters equal to or less than 2.5 μm ($PM_{2.5}$), the causative substances and sources of $PM_{2.5}$ must be clarified. In Japan, not only the influence of transboundary pollution caused by advection from overseas but also the contribution of domestic sources may be significant when $PM_{2.5}$ concentration rises. Therefore, the influence of domestic sources such as automobiles, ships, outdoor incineration, weather conditions, etc., must be considered. In the following, epidemiological study on the effects of $PM_{2.5}$ and ozone (O_3) to asthmatic attacks in an urban area, Himeji city, in Japan is explained based on the literature [18,19]. The summary of the research is given below as (1) and (2):

1. The primary chemical components of $PM_{2.5}$ in the atmosphere and coarse particles are continuously measured, and the associations with asthmatic attacks are evaluated every week and every day to reveal the particle components affecting asthma.
2. The difference in the $PM_{2.5}$ and O_3 sources, the effect on the respiratory system considering the source analysis, and the effects of transboundary pollution are clarified.

The relationship between the weekly mean value of particle size–dependent air pollutant component concentrations and the number of patients with asthma attacks in the following week were analyzed by adjusting the influence of atmospheric pressure, temperature, humidity, sunshine time, season, and year. In addition to the analysis throughout the year, analysis by age and season was conducted. The end of the year, the beginning of the year, April–May, the consecutive holidays in September, and the summer vacation were excluded from the analysis because the number of asthma attacks patients may not be accurately counted during these periods. The summarized research results are as follows:

1. As the concentration of $PM_{2.5}$ in the atmosphere increased, the number of patients with asthma attacks significantly increased.
2. The increases in the concentrations of water-soluble organic compound (WSOC) and H^+ in $PM_{2.5}$ and WSOC in PM_{10}–$PM_{2.5}$ had significant effect on the increase in the number of patients with asthma attacks.

3. By season, the effect of WSOC and NO_3^- in $PM_{2.5}$ in spring and H^+ in summer on asthma attacks was significant.
4. Asthma attacks increased significantly with an increase in atmospheric O_3 concentration. Especially, the correlation was significant in the summer.

On the other hand, to know the association between the chemical components of $PM_{2.5}$ and night-time primary care visits (PCVs) due to asthma attacks of children, a case-crossover study was performed [19]. The subjects were 1251 children aged 0–14 years who received medical care for asthma at a municipal emergency clinic. Daily average concentrations of the hydrogen ion, sulfate ion, nitrate ion, and WSOCs, which are components of $PM_{2.5}$, were measured. The odds ratios (ORs) of PCVs per unit increment (interquartile ranges [IQRs]) in each chemical component of $PM_{2.5}$ were estimated for the subgroups of warmer months and colder months separately.

The measurement was performed in Himeji city, Japan, by Hyogo College of Medicine for the major chemical components in sulfate ions (SO_4^{2-}), nitrate ion (NO_3^-), WSOCs, and hydrogen ion (H^+) in $PM_{2.5}$. This study was carried out at the Himeji City Emergency Clinic. This facility was set up to provide emergency care between 9 p.m. and 6 a.m. on weekdays. Himeji is located in the western part of Japan facing the Seto Inland Sea about 100 km west of Osaka. Land area is 534 km, and the estimated population is 530,000 in 2017, of whom 74,000 are estimated to be aged less than 15 years.

Table 5.4.2 shows the association between $PM_{2.5}$ with its components and PCVs with asthma attacks for children under 14 years old. The effects of $PM_{2.5}$ mass concentration, SO_4^{2-}, NO_3^-, WSOCs, and H^+ on relative risk of asthma attacks were recorded. ORs per unit increment of $PM_{2.5}$ and each chemical component of $PM_{2.5}$ were estimated for the subgroups of warmer (May to October) and colder (November to April) months ($N = 817$ and 434, respectively) in 2014. The unit increments in the analysis were IQRs of $PM_{2.5}$ mass concentration, sulfate ions, nitrate ions, WSOCs, and hydrogen ions (11.0 $\mu g/m^3$, 26.9 $nmol/m^3$, 16.3 $nmol/m^3$, 0.69 $\mu g\ C/m^3$, and 5.30 $nmol/m^3$, respectively). With regard to the time lag, the association between daily average air pollutant concentration (with the day of PCV [L0] and the previous day [L1], respectively) and night-time PCVs due to an asthma attacks was analyzed. The association between the three-day average concentration of air pollutants (L0–2) and PCVs was also

Table 5.4.2 Point-estimated odds ratio[b] of the association between $PM_{2.5}$ components and PCVs with asthma attacks for children under 15 years at the clinic [19].

	$PM_{2.5}$ mass concentration			SO_4^{2-}			NO_3^-			WSOCs			H^+		
	L0	L1	L0–2	L0	L1	L0–2	L0	L1	L0–2	L0	L1	L0–2	L0	L1	L0–2
Warmer months (May to October) (N = 817)	1.08 (+)	1.08 (+)	0.97 (−)	0.92 (−)	0.85[a] (−)	0.87 (−)	1.39 (+)	0.76 (−)	0.89 (−)	0.94 (−)	1.33[a] (+)	1.13 (+)	1.05 (+)	1.18[a] (+)	1.06 (+)
Colder months (November to April) (N = 434)	0.95 (−)	1.02 (+)	1.02 (+)	1.01 (+)	0.83 (−)	0.90 (−)	1.13 (+)	1.00 (+)	0.81 (−)	0.75 (−)	1.27 (+)	1.19 (+)	1.08 (+)	0.92 (−)	1.14 (+)
Degree of agreement[c]		c			c	c	c		c	c	c	c	c		c

[a] $p < 0.05$.
[b] Odds ratios chosen as point estimates were marked "+" if they exceeded 1 and "−" if they were less than 1.
[c] Columns are marked "c" if "+" and "−" for items which coincided between colder and warmer months.

analyzed. The degree of agreement between the warmer and colder months was analyzed. ORs were chosen as point estimates and marked "+" if they exceeded 1 and "−" if they are less than 1 in the table. On L0−2, the ORs exceeded one for all four components of $PM_{2.5}$. On both L0 and L1, the ORs exceeded 1 for three out of four components.

In the results, no association was seen between PCVs and $PM_{2.5}$ mass concentrations the day before the PCVs in either warmer or colder months. In the warmer months, an association was seen with the concentrations of WSOCs and hydrogen ion the day before the PCVs (OR = 1.33; 95% confidential interval (CI): 1.00−1.76, OR = 1.18; 95% CI: 1.02−1.36, respectively). Furthermore, a negative association was seen between sulfate ion and PCVs (OR = 0.85; 95% CI: 0.74−0.98). No associations were observed in the colder months. In conclusions, it was observed that a positive association between PVCs and certain concentrations of WSOCs and hydrogen ions in the warmer months. In contrast, sulfate ion showed a negative association.

5.4.2 Relationship between emissions regulation and particle concentration from ships

Fig. 5.4.3 shows a schematic diagram of the distillation process from crude oil to various fuel products such as liquefied petroleum gas, gasoline, naphtha, kerosene, jet fuel, light oil, heavy oil, bunker oil, lubricant oil, asphalt, etc., including the ship fuels [20]. In the distillation column, crude oil is distilled by heating, and it is separated into various fuels by temperature difference. In ships, a small amount of gas oil is added to low-quality heavy oil or bunker oil and is then shipped after adjusting the sulfur content and kinematic viscosity.

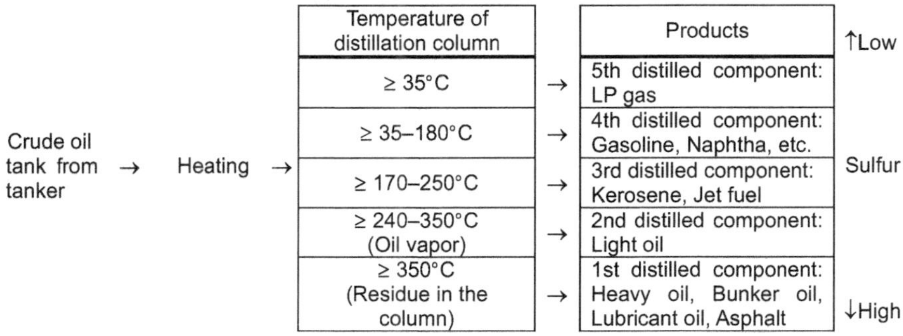

Figure 5.4.3 Distillation process from crude oil to various fuel products including the ship fuels. *(Made based on the figure in https://www.ekka.co.jp/business/airplane.html.)*

The reason why the sulfur content regulation of fuel oil becomes the PM emission countermeasure is as follows: On investigating the emission factor in ship exhaust gas based on the dilution method, it is observed that the exhaust gas is discharged in the state where multiple components are condensed. The emission intensity of sulfuric acid mist, which is considered as particulate, increases with an increase in sulfur content in fuel. Therefore, the PM regulation is mainly targeted to sulfuric acid mist induced by sulfur in fuels.

On the other hand, the emission intensity of particulate carbonaceous species basically does not depend on the sulfur content of fuel oil. Therefore, another framework of regulation is necessary. Especially, black carbon (BC) deposition in the Arctic Circle by the ship exhaust gas recently becomes an environmental problem. The utilization of the Arctic route is being debated owing to the decrease in sea ice owing to not only global warming but also additional melting by BC deposition on ice. Albedo which is defined as the ratio of the reflected solar radiation to the incident one decreases owing to deposition of BC on ice-sheet by ship's exhaust gas. Furthermore, concern about the effect of radiation enforcement in the area becomes high. Consequently, short-lived climate pollutants (SLCPs) become a problem. SLCPs are chemical substances that have relatively short chemical longevity in the atmosphere of several days to several tens of years and are considered substances that have an action to change the climate. Major substances of SLCPs are methane, tropospheric ozone, and BC particles, which are major components of the soot generated when burning fossil fuels and biomass; undoubtedly, these are warming the atmosphere and are simultaneously air pollutants that negatively affect human health and life, agriculture, and ecology [21]. Therefore, BC particles from ships should be reduced further as well as NO_x and SO_x. Effective aftertreatment method for carbon PM is explained in Chapters 3 and 4.

5.4.3 Examination and status of emission control area setting in Japan

By setting the ECA for exhaust gas components NO_x and SO_x from ships, the environment in the relevant area is greatly improved. However, the setting has a negative economic impact, and its setting should be done with caution. The ECA has already been set up in the Baltic Sea, the United States, etc., as explained in Chapter 2. In this subsection, we will introduce the examination and status of the recent ECA setting in Japan [20]. A Japanese ECA technical committee was established as the secretariat of the Ministry of Land, Infrastructure, Transport, and Tourism. It was judged in 2012 by the simulation analysis that the ECA was unnecessary in Japan. The following reasons are described.

1. SO_2 and NO_2 will be expected to meet environmental standards in the future by continuing the existing onshore emission control methods.
2. Even if ECA for SO_x (0.1% sulfur in ship fuels) is set, which is more stringent than the global regulation (0.5% sulfur in ship fuels) starting from 2020, it is estimated to have a limited effect on the increase of atmospheric $PM_{2.5}$ concentration.
3. Even if ECA for NO_x (approximately 80% NO_x reduction) is set, which is more stringent than the global regulation (approximately 20% NO_x reduction), it is estimated to have a limited effect on the atmospheric photochemical oxidant's concentrations.
4. Negative economic impact of the ECA setting is high. A large cost of low-sulfur fuel preparation, an inverse modal shift from ship to automobiles, and decreases in sightseeing ferries may occur.

Generally, in considering the introduction of environmental regulations, the following three items are to be considered:

a. The expected concentration reduction effect by introducing the regulation should be estimated.
b. The insurance money and medical expenses that can be reduced based on the decrease in the number of deceased people and the number of hospitalized patients expected by introducing the regulation should be estimated.
c. The cost required for capital investment for the emission control system necessary for introducing the regulations, which are considered to be internal cost, should be estimated.

In Europe and the United States of America, the ECAs were set, and in China, it is planned to be set up in ports of large cities such as Shanghai city. For the decision of the ECA setting on the US coast, the health impact assessment of b was important. In the United States of America, a, b, and c were considered to decide whether the ECA is to be introduced or not, while in the case of Japan, a and c were mainly considered.

5.5 Reduction of CO_2 emissions from marine engines

5.5.1 Methods to reduce CO_2 from main engine

Carbon dioxide (CO_2) reduction should be considered for the entire ship or maritime transportation. In this subsection, the CO_2 reduction attempt from the main propulsion diesel engine is explained. Firstly, the energy efficiency design index (EEDI) that expresses CO_2 emission amount from a diesel engine is conceptually defined in Eq. (5.4) [22].

$$\text{EEDI}(\text{g}(\text{CO}_2)/\text{ton}/\text{mile}) = \frac{\text{Engine power (kW)} \times \text{Fuel consumption rate (g/kWh)} \times C_F}{\text{DWT (ton)} \times \text{Speed (mile/h)}} \tag{5.4}$$

where C_F is the CO_2 conversion coefficient. It is noted that Eq. (5.4) is a simplified one, and the detailed formula of EEDI is provided by the IMO. Based on the equation, the method to reduce EEDI is described. The targeted new regulation is explained in A.2 of Appendix. The following five methods are considered to reduce EEDI from the engine side [22].

1. The maximum ship speed is set to a lower value and adopt new main engines with lower power ones.
2. The propulsion efficiency is improved by combining lower rotational speed and large diameter propellers, and developing new main engines with low-rated speeds.
3. The fuel consumption rate of the main engine or generators is reduced by recovering the waste heat of the main engine.
4. The thermal efficiency of the main engine unit itself is improved and the fuel consumption rate is reduced.
5. Fuel is switched from heavy oil to natural gas, hydrogen, or biofuel to decrease CO_2 emission.

In the method 1, some comments are described. The engine power is proportional to the cube of the ship moving speed. Therefore, when the engine power is reduced by 50% in Eq. (5.4) on EEDI, the ship's moving speed is reduced only by approximately 80%. As a result, EEDI decreases by $80-50=30\%$, and it can comply with the new regulation of ships after 2025 as explained later in Table 5.5.1. However, limitation exists on the real ship speed. In the method 2, lower rotational speed and large diameter propeller can decrease EEDI from the viewpoint of propulsion efficiency increase. Many trials have been carried out from the viewpoint of this such as the main engines with lower-rated speeds.

In the method 3, Fig. 5.5.1 shows the thermal balance of the engine, which expresses the ratio or percentage of the heat generation quantity of the fuel to the shaft output of the marine engine. This value is called the net thermal efficiency, and the fuel consumption rate is inversely proportional to this value, which is known from the definition. As shown in the figure, the thermal efficiency at the rated load of the low-speed two-cycle engine is approximately 50%. However, it is difficult to realize further improvement.

Table 5.5.1 Reduction phases of required energy efficiency design index (EEDI) reduction rate [23].

	Year	EEDI reduction rate %
Phase 0	2013–14	0
Phase 1	2015–19	−10
Phase 2	2020–24	−20
Phase 3	2025–	−30

According to the second law of thermodynamics, the thermal efficiency of 50% is the highest level in the heat engine. Fuel consumption during partial load, which is often used in the actual operation of marine diesel engines, has improved considerably using the recent electronic engine control method. In the figure, approximately 50% of the heat generation by the fuel is converted into the shaft output, and most of the remaining portion is released as the exhaust heat. The exhaust gas (25.5%) in the figure corresponds to the calorific value of the exhaust. A method to be considered is to effectively

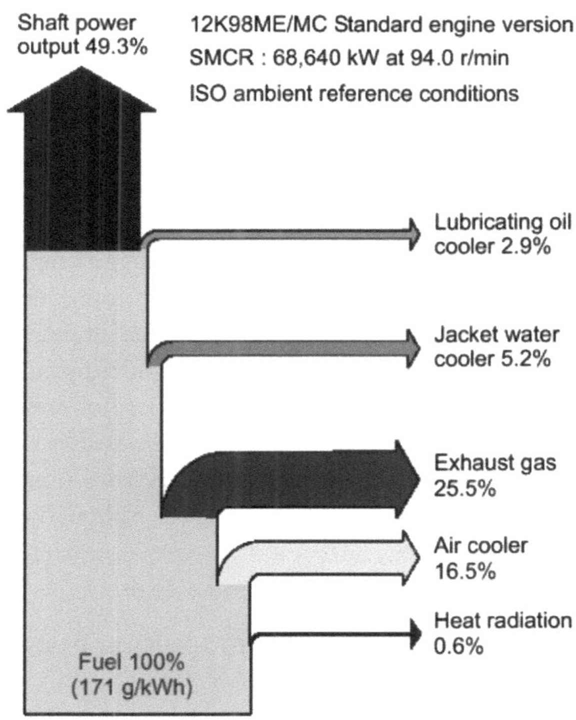

Figure 5.5.1 Thermal balance of low-speed two-cycle main propulsion diesel engine [22].

recover this because other losses such as consumptions in lubricating oil cooler, jacket water cooler, air cooler, and heat radiation are difficult to reduce for the engine operation. The air cooler (16.5%) in the figure is the amount of heat of the turbocharged air cooler. One example is shipboard power generation with exhaust gas waste heat boilers plus steam turbine. Recently, the research and development to increase waste heat recovery is thriving. CO_2 reduction by such systems is also counted for the reduction of EEDI. Exhaust turbines of the turbocharger certainly use the exhaust energy; however, it has been used only for compressing the air entering the engine. As a result, it does not lead to raise the thermal efficiency directly. Recently, a method of energizing the crankshaft of the main engines with exhaust gas energy directly has been developed as explained in Subsection 5.5.2.

The EEDI and ship energy efficiency management plan were adopted by the IMO to reduce CO_2 emissions from international shipping, which is a partial amendment to the MARPOL 73/78 convention Annex VI. This became effective on January 1, 2013 [23]. Table 5.5.1 shows the reduction phases of the required EEDI reduction rate in the future. By combining both methods on the ship structure side and the engine side, the EEDI regulation must be complied, in which a 20% decrease starting from 2020 and a 30% decrease starting from 2025 are expected based on the IMO regulation. If the CO_2 reduction target reaches 50–80% in the future, hydrogen processed from renewable energy or biofuels must be adopted. Marine fuel cells are also an alternative power source in the future.

Currently, although energy-saving technology is primarily focused on ship body improvement, the efficiency of marine engines must be improved and renewable energy and alternative energy are to be used. Trends in greenhouse gas (GHG) regulations of international shipping and the use other than conventional fuels lead to CO_2 emission reductions. Although a lengthy time is required, the shift to fuel that does not emit CO_2 will be further popular. However, reducing GHG emissions will be more important by improving the fuel efficiency of marine engines when using conventional fuels such as heavy oil during the shifting process. Although marine engines are more efficient than others, further improvement efforts in efficiency and the innovation of marine diesel engines are necessary.

5.5.2 Exhaust heat recovery with propulsion assistance for main engine

Exhaust heat recovery system with propulsion assistance for main engines effectively utilizes the waste thermal energy contained in the exhaust gas of

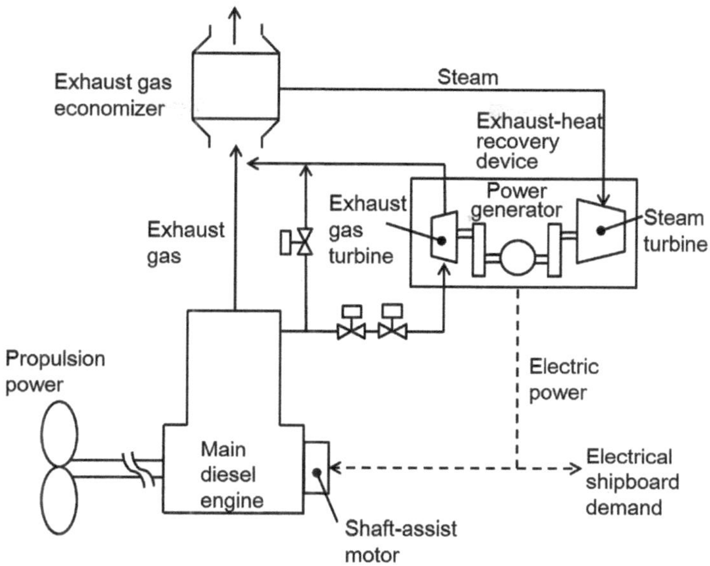

Figure 5.5.2 Configuration of the exhaust heat recovery system [24].

the main engine [24]. Fig. 5.5.2 shows a configuration of the exhaust heat recovery system. In this system, electricity is generated utilizing the thermal energy contained in the exhaust gas of the main engine, ultimately by the exhaust heat recovery device that combines the steam turbine and the exhaust gas turbine. As shown in the figure, bypassed exhaust gas drives the exhaust gas turbine, and steam generated by the exhaust gas economizer drives the steam turbine. Electric power is generated by the power generator connected to the two turbines. The electric power obtained covers the electrical shipboard demand, and the surplus electricity can be reused to a shaft—assist motor mounted on the main shaft of the main engine, thereby assisting the propulsion of the ship. Hitherto, there have been cases in which ships require large electric shipboard demands such as a container ship carrying refrigerated containers with reused waste heat and covered shipboard electricity.

Nevertheless, for ships with low electrical shipboard demand, such as crude oil tankers and ore shipping carriers, because all the electricity that can be recovered from waste heat was not able to be used, much exhaust gas energy is abandoned. A system that assists propulsion using the shaft—assist motor developed as shown here is an advanced attempt in that the exhaust heat recovery, regardless of the ship type that can be effectively used.

Table 5.5.2 Main specifications of the exhaust heat recovery system [24].

Ship specification	Total length	Approximately 330 m
	Width	Approximately 57 m
	Deadweight	Approximately 250,000 ton
Engine plant	Main engine	MAN B&W 7S80MC-C Mark7 21,910 kW × 74.5 rpm
	Exhaust heat recovery device	1700 kW × 1800 rpm
	Shaft—assist motor	Motor output 720 kW Generator output 250 kW
	Diesel generator	960 kW × 900 rpm × 3 sets

Table 5.5.2 shows the details of the ship and engine plant equipped with the exhaust heat recovery system.

The amount of electric power generated by the exhaust heat recovery device was confirmed on the sea with this system installed on the actual ship. In a result, as the load on the main engine became higher, the amount of electric power that can be recovered using the exhaust heat recovery device tended to be larger. In the result, approximately 650 kW of electric power was used inside the ship regardless of the load on the main engine; it is usually supplied by diesel generators in ordinary ships. In the ship, by using the generated electric power from the main engine, the diesel generator could be stopped and the fuel consumption could be reduced when the main engine load became approximately 60% or higher. The surplus electricity can be reused to a shaft—assist motor that assists the propulsion of the ship. Approximately 7% of the higher fuel consumption reduction was recorded for 60% of the main engine load, and it became approximately 8% for 85% of the main engine load.

5.5.3 Possibility of improvement on marine diesel engine thermal efficiency

In the method 4 in Subsection 5.5.1, the low-speed two-cycle diesel engine for large-scale marine engines has been greatly improved in a fierce fuel consumption rate competition. Consequently, the thermal efficiency is highest in a single engine as shown in Fig. 5.5.3. This high thermal efficiency that is almost the same as that of gas turbine combined cycle power generation is supported by several factors such as large cylinder diameter, long stroke, high air excess ratio, low speed, etc. It is considered that various losses are reduced, and a thermodynamic cycle closer to the

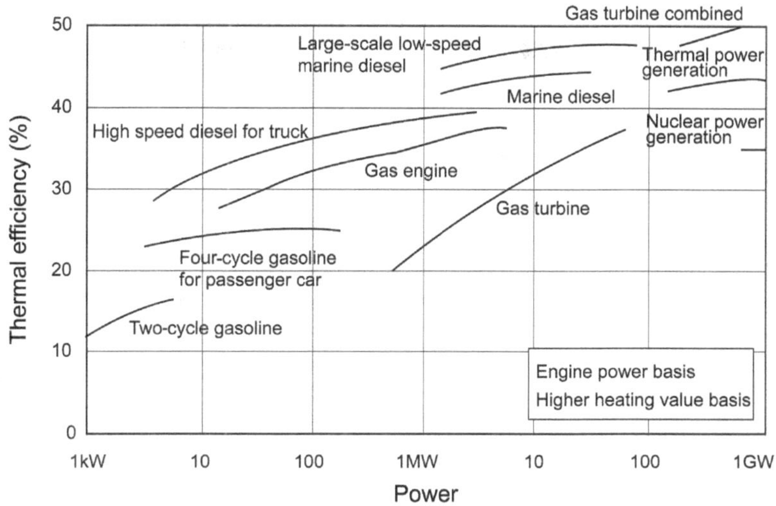

Figure 5.5.3 Thermal efficiencies of various engines [25].

ideal one is realized. Hereafter, further improvements will be desired to perform CO_2 emission reduction. The following targets can be considered [25]: optimization of a turbocharger system such as improvement in turbocharger efficiency of a single unit, improvement in pressure ratio, introduction of two-stage turbocharging system, increase in excess ratio due to the scavenging efficiency improvement including the optimization of scavenging port, decrease in mechanical loss and heat loss, and increase in pressure ratio P_{max}/P_s, where P_{max} is the highest pressure in the cylinder and P_s is the scavenging pressure or turbocharging pressure.

The shipping is the most ecological transportation mode. The 747 aircraft of Boeing Company emits 435 g/(ton·km) of CO_2. On the contrary, the emission from a large container ship (18,000 TEU) is only 3 g/(ton·km), which is approximately 1/150 of that of the 747 [26]. Cargo owners can reduce the CO_2 emissions related to the transport of their products by accelerating the modal shift from aviation and rail to shipping. In addition, environmental concerns are also rising, such as the data disclosure of the ship's CO_2 and air pollutants by NO_x and SO_x emissions become a condition in bidding for transportation business. Additionally, the development of energy-saving devices and the approach toward optimum route search are progressing.

5.5.4 CO_2 reduction by fuel shift from heavy oil to natural gas

In the method 5 in Subsection 5.5.1, fuel change from heavy oil to natural gas is explained. The typical component of MDO is represented by hexadecane or cetane $C_{16}H_{34}$ (molar mass = 226.50 g/mol), while the main component of natural gas is methane CH_4 (molar mass = 16.05 g/mol). These are burned to generate the heat quantity according to the following reactions:

$$C_{16}H_{34} + 24.5O_2 \rightarrow 16CO_2 + 17H_2O(l)\ \Delta H_1 = -10,700\ \text{kJ/mol} \quad (5.5)$$

$$CH_4 + 2O_2 \rightarrow CO_2 + 2H_2O(l)\ \Delta H_2 = -890.3\ \text{kJ/mol} \quad (5.6)$$

where ΔH_1 and ΔH_2 are the standard enthalpies of combustion of cetane and methane, respectively, and (l) means liquid. The standard enthalpy of combustion equals to the higher heating value of fuel. Because methane has a lower C/H ratio than cetane, the mass of CO_2 (molar mass = 44.01 g/mol) emission per unit mass of methane (= 2.742 g/g) becomes less than that of cetane (= 3.109 g/g). In the fuel change from the diesel oil to natural gas, the CO_2 conversion factor, C_F, decreases in Eq. (5.4) for the EEDI and becomes 2.742/3.109 = 0.88 times. Further, as natural gas produces higher heat per unit mass than the diesel oil, for the same thermal efficiency, the fuel consumption rate per unit combustion enthalpy of methane (= 18.03 mg/kJ) is lower than that of cetane (= 21.17 mg/kJ), that is, 18.03/21.17 = 0.85 times lower. As a product of the two values, the EEDI decreases approximately $0.88 \times 0.85 = 0.75$ times and may clear the target value of a 20% decrease to be imposed on new shipbuilding after 2020 [22].

Additionally, because natural gas is a clean fuel that does not contain sulfur and nitrogen, it is also suitable for NO_x and SO_x regulations, which are now imposed in the world as explained previously. In particular, compared with heavy oil, it is easier not only for a four-cycle medium-speed engine (output power = several thousand kW) but also for the low-speed main engines (output power = several tens of thousands kW) to maintain the NO_x regulations in the ECA. Further, the emissions of smoke (BC) and particulate matter (PM) are extremely small. The future popularization depends on the price of natural gas. Further, after the global cap enactment of fuel with less than 0.5% sulfur by IMO in 2020, either MGO or MDO will be a price competitor for gas fuel instead of HFO.

5.5.5 Reduction of greenhouse gas using biofuels and other fuels

In the method 5 in Subsection 5.5.1, further reduction of GHG using biofuels [27,28] is explained. When biofuels made from plants are burned, it is considered the total amount of CO_2 in the atmosphere does not increase because the next-generation plants grow by absorbing it through photosynthesis. Therefore, it is decided that CO_2 emitted by burning biofuels is not counted as GHG emissions. From this consideration of carbon neutral, biofuels are considered as one way of reducing GHG emissions and a renewable energy source. Nevertheless, to reduce GHG emissions, improving the efficiency of combustion engines is also another direction. However, it is difficult as marine diesel engines are already highly efficient. It is challenging to manufacture an engine that reduces 30% of the fuel consumption to obtain the same output. However, as the same effect is obtained by mixing 30% of biofuel, it is considered a more practical countermeasure from the viewpoint of CO_2 emission.

In addition to direct burning, biomass fuels are converted or processed into liquid, gaseous, and solid fuels that meet the usage purpose. Among them, liquid fuel is used as a substitute for gasoline or diesel oil and is the strongest contender as a fuel for vehicles. In ships, alternative sources of fuel for diesel oil are considered promising, as most of the power sources for ships are diesel engines. To use oils for marine diesel engines, low viscosity is necessary. Three ways can be used to decrease the viscosity: (1) heating, (2) mixing with diesel oil, and (3) esterification. When raw vegetable oil and methanol are subjected to a transesterification reaction in the presence of a catalyst, a methyl-esterified product of higher fatty acid is produced. This is called biodiesel fuel (BDF). More recently, technological developments for the next-generation BDF of hydrotreated vegetable oil (HVO) and bio to liquid (BTL) have been conducted. However, in marine diesel engines, because even HFO with high viscosity is used by heat addition, raw vegetable oil with high viscosity that is not esterified may be used. Table 5.5.3 presents the oil-producing plants considered as alternative fuels for diesel oil [27]. Representative oil-producing plants (oil name) such as palm (palm oil), soybean (soybean oil), rapeseed (rapeseed oil), and sunflower (sunflower oil) are popular. In particular, palm produces the highest oil per unit area and is promising from the viewpoint of biofuel production. Because the average oil

Table 5.5.3 Representative oil-producing plants [27].

	Oilseeds fruit yield per unit area ton/ha	Average oil content %	Fat and oil yield per unit area ton/ha
Palm (palm oil)	20	20	4.00
Soy (soybean oil)	1.8–2.3	17	0.3–0.4
Rapeseed (rapeseed oil)	1.4–2.5	35	0.5–0.9
Sunflower (sunflower oil)	1.4–2.3	35	0.5–0.8

content and fat and oil yield per unit area is relatively high, the cultivation amount is rapidly increasing especially in Southeast Asia. It is noted that to reduce CO_2 emissions, the uses of fuel cell and hydrogen fuel engine are alternative methods in the future.

5.6 Past, current, and future markets for aftertreatments

In this section, the past, current, and future markets for aftertreatments are described. The exhaust system market for vehicles including passenger cars, truck, bus, etc., and ships is projected to grow at a compound average growth rate of 8.45% to reach a market size of 59.02 billion USD by 2021 [29]. The major driving factors for the exhaust system market are the increase in vehicle and ship production and the stringency in the emission regulations. With an increase in the number of electric passenger cars, the market for automobile aftertreatment systems has unknown factors. However, larger combustion engines for marine ships have a long history, will not be replaced easily, and will last for a long time. Therefore, aftertreatment apparatus is indispensable. For ships, the following recent emission aftertreatment apparatus will be open in the market. The market sizes for the apparatus are estimated in the following subsections.

5.6.1 Implementation cost of ballast water management system

Ballast water is the water used in the ballast of a ship, which is a kind of weight loaded on the bottom of the ship. When a cargo ship leaves a port without cargo, seawater is loaded at the port and discharged outside the ship at the other port where cargo is loaded. Because the contained aquatic organisms in the ballast water affect the regional ecosystem, regulations under international treaties are in progress. Therefore, in September 2017, the IMO decided that from September 2017, all ship-owning companies must comply with the requirement that a ballast water management system (BWMS) needs to be installed on each ship. An investment cost amount of up to 2.5 million USD per ship is required [30]. In another reference, the price of the BWMS is reported as 40–200 million Japanese Yen (=350 thousands–1.8 million USD) [31].

5.6.2 Implementation cost of SO_x emission control system

In October 2016, the IMO has decided to reduce the global permissible sulfur oxide (SO_x) emission by 2020, which requires a reduction in the sulfur content of marine fuel from 3.5% to 0.5% (0.1% in the ECA in 2015, ECA for SO_x: the Baltic Sea area, the North Sea area, the North American area, and the Caribbean Sea area) or the installation of exhaust gas cleaning system (EGCS) such as marine diesel engine scrubber. Using an LNG gas engine is also another option but requires extensive changes to the existing ships; the cost of the SO_x scrubber is estimated as 3–6 million USD [30]. For burning the high-cost low-sulfur fuel, the incremental cost is estimated currently as 150 USD per ton. Assuming 30 metric ton per day of fuel consumption, the scrubber investment costs would amortize over up to 4 years for many ships.

As an example of a commercialized product, Fuji Electric Co., Ltd. chemical scrubber [32] is announced. It achieves a 0.1% regulated value and contributes to the suppression of cost. Its features characterize the industrial and compact design. The market size of the exhaust gas cleaning system is estimated to be approximately 50 billion Yen (= 440 million USD) in 2017. After strengthening the regulations, it is necessary to select either an expensive low-sulfur fuel or the scrubber system. By installing the system, the use of the current fuel (HFO) can be continued. It is estimated that the investment collection period of this apparatus is less than 3 years. It is noted that this estimation was conducted based on the following assumption: As of March 2014, the fuel consumption of bulk cargo ship is 95,000 DWT (deadweight tonnage). This cyclone type chemical scrubber achieves approximately 50% downsizing compared with conventional equipment, and it is very suitable for marine diesel engines because the additional space inside the ship is very restricted.

5.6.3 Implementation cost of NO_x emission control system

NO_x emission from marine engines is a great problem, and ships entering the ECA should be equipped with procedures that decrease NO_x emission. SCR is a very effective method, and the estimated cost of the SCR system is 30 USD/kW [10]. The market size for the aftertreatment systems can be estimated in the next subsection.

5.6.4 Market size of aftertreatment systems

Based on the description above, the market sizes for the aftertreatment systems of ballast water treatment apparatus, scrubber for SO_x reduction, and SCR for NO_x reduction are estimated. Table 5.6.1 summarizes the estimation results above on the market size for the aftertreatment apparatus. The target is the number of ship completions in the world as reported in Ref. [33] by the Shipbuilders' Association of Japan. For example, in 2016, the world total number is 2543 (ship size coverage: 100 gross tonnage (GT) and over). The assumption for the installation rates of each aftertreatment apparatus is as follows: BWMS must be installed after 2016, but not before 2015. However, because the installation has started since around 2011, it is

Table 5.6.1 Estimation results on market sizes of aftertreatment apparatus for ship size coverage of 100 GT and over.

Year	2013	2014	2015	2016	2017 1Q
Japan	540	522	520	514	164
South Korea	386	343	358	359	95
China	1073	914	949	824	236
Europe	159	166	152	165	46
Others	931	1018	891	681	130
World total	3089	2963	2870	2543	671
Implementation rate of BWMS (1.1 million USD for a ship)	60%	80%	90%	100%	100%
World market size of BWMS, billion USD	2.0	2.6	2.8	2.8	0.7
Implementation rate of scrubber (4.5 million USD for a ship)	10%	20%	30%	40%	50%
World market size of scrubber, billion USD	1.4	2.7	3.9	4.6	1.5
Implementation rate of SCR (5 million USD for a ship)	10%	20%	30%	50%	50%
World market size of SCR, billion USD	1.5	3.0	4.3	6.4	1.7

BWMS, ballast water management system.

therefore assumed that the mounting rate will be increased stepwise. The average price of the product is assumed to be 1.1 million USD for a ship in this estimation. EGCS for SO_x or the low-sulfur fuel is obliged to be used after 2015 for ships entering the ECA. After 2020, it should be used for all ships. It appears that the installation has started since around 2013. It is assumed that 50% of the ships select the scrubber rather than the low-sulfur oil fuel in 2017. The average price of the product is assumed to be 4.5 million USD for a ship in this estimation. A NO_x emission control apparatus or a De-NO_x system is to be used after 2016 for ships entering the ECA. It is assumed that 50% of the big ships are equipped with the SCR apparatus. The average price of the product is assumed to be 5 million USD for a ship in this estimation.

The cost for the LNG ship is described in Subsection 5.2.3. The market size is calculated similarly as explained in this subsection. As shown in the table, in 2016, the world market size of BWMS is estimated as 2.8 billion USD, that of the scrubber is estimated as 4.6 billion USD, and that of the SCR is estimated as 6.4 billion USD. In the near future, the market size will continue with the similar scale.

5.7 Operation and maintenance of aftertreatment systems

The maintenance of the aftertreatment system for a diesel engine is a very important issue because the emission control should be performed well during the whole life cycle of the engines [34]. For example, on the NO_x emission control, whether or not the system conforms to the regulation is judged by an engine manufacturer under inspection by a designated inspection agency. When it is passed, the engine manufacturer will attach the certification letter to each engine and send it to the shipyard company. The components of the exhaust gas of the diesel engine will vary depending on the explosion pressure of the engine, the combustion temperature, the compression ratio, the situation of fuel injection, the shape of the combustion chamber, etc. After the engine is attached to the ship, NO_x emissions change if these operation parameters are changed. To prevent this, the handling manual (technical file) of the engines is attached to the certification sheet, and it is obliged to keep it on the ship. During midterm and regular ship inspection, it is confirmed by the inspector of the Transportation Bureau based on the handling manual of the engines whether the engine is kept in the same state as the delivery without any refinement. Therefore, the operator has to maintain the engine constantly on the street. As for engine parts replacement, this inspection will not be passed unless genuine parts with manufacturer's inscription are used. The explanations for detailed operation and maintenance to keep the constant performance of the aftertreatment system are presented in the following subsections.

5.7.1 Operation and maintenance of NO_x reduction systems

In ships, EGR and SCR are the main NO_x reduction technologies. For example, EGR and SCR technologies are developed by several companies. In general, EGR and SCR do not differ much in terms of the life cycle cost which include operation and maintenance [35]. The initial and running costs for 20 years are also almost the same. The operation and maintenance of the EGR and SCR are explained below.

Operation and maintenance of exhaust gas recirculation

EGR is one of the reduction technologies for NO_x emission, as explained in this book. In comparison with SCR, because the exhaust gas temperature is not limited, EGR can be operated even under the condition of a low engine load and low exhaust gas temperature. However, because the circulated exhaust gas includes diesel particulates and SO_x, engine problems such as the corrosion of the combustion chamber and exhaust pipe, abnormal wear of rings, and degradation of cylinder oil could occur. As a solution to these problems, particularly in two-cycle marine diesel engines, a scrubber can be used for purifying circulated exhaust gas before it is used in the engine intake. When a scrubber is used, an additional maintenance step is required, in that the wasted scrubbing water should be treated as required. Because the water loaded in a ship is limited, the water should be reused. As an example of the wasted scrubbing water treatment in a low-pressure loop EGR system (Ref. [27] in Chapter 4), the wastewater is separated into the treated water and soot by centrifugal separators. The separated soot is removed as sludge. The treated water is reused after a pH control. The sludge should be unloaded as industrial waste because ocean disposal and incineration on the ship cannot be conducted because of restrictions. In this unloading process, the sludge is warmed at 35°C to obtain the fluidity for pumping. For the stable operation of EGR, these maintenances and additional processes are required.

Operation and maintenance of selective catalytic reduction

SCR is also another reduction technologies of NO_x emission, as explained in this book. In this technology, the NO_x in the exhaust gas is reduced to N_2 and H_2O by the selective reaction of NO_x and reductant ammonia (NH_3) on a catalyst. However, NH_3 is difficult to handle and careful management is required. Hence, urea aqueous solution is loaded on a ship and NH_3 is obtained from the urea aqueous by hydrolysis at a high temperature. In the operation, the urea aqueous is supplied to the exhaust gas flow pipe by injection. In the operation of the SCR system, urea aqueous solution storage and supply apparatus are necessary. This requires additional space on a ship and running cost. On the other hand, the emission and leakage of unreacted NH_3 is called the ammonia slip. The ammonia slip is also a problem and should be prevented. To prevent the ammonia slip, the appropriate control of urea aqueous injection and the sufficient volume of catalyst are required. The reaction of sulfur in fuel and NH_3 forms ammonium sulfate ($(NH_4)_2SO_4$) and ammonium hydrogen sulfate (NH_4HSO_4). The generated NH_4HSO_4 covers the surface of the catalyst. In this situation, the performance of the reaction in the SCR process is degraded. NH_4HSO_4 is easily generated as a byproduct under the condition that the concentration of sulfur is high and the exhaust gas temperature is low. The exhaust gas temperature should be maintained sufficiently high (> 300°C) for the reactions of hydrolysis and NO_x reduction.

In a bypassed test operation of SCR for a marine diesel engine (4S50ME-T9, Hitachi-MAN B&W), corrosion and adhering substances are observed in the vaporizer, SCR reactor, and catalyst. The corrosion and adhering substances are thought to be caused by sulfur in the fuel (3.18 mass%) and the low temperature of the exhaust gas. In addition, adhering substances are also observed inside an exhaust gas economizer installed downstream of the turbine. These adhering substances cause rapid pressure loss in the economizer. These adhering substances are NH_4HSO_4, and they are observed while the ammonia slip occurs. The relationship between pressure loss in the economizer and ammonia slip is revealed with different ammonia slip conditions of 80 ppm, 10 ppm, and 5 ppm (Ref. [7] in Chapter 4).

In summary, the exhaust gas temperature should be maintained sufficiently high in the operation of SCR. Low-sulfur fuel is preferable. The accumulating or adhering substances on the catalyst degrade the SCR reactions; in other words, the system performance is reduced. In addition, this kind of problem can occur in other devices such as the exhaust gas economizer. During maintenance, the accumulating or adhering substances should be removed. The substances are $(NH_4)_2SO_4$, NH_4HSO_4, and soot. Some physical removals or heating could be applied. If the catalyst is damaged, it should be replaced by a new one.

The marine SCR system has been developed and has reached a commercial level from the viewpoint of emission control. However, some practical problems remain. One is the problem of space. Because the space on a ship is limited, the apparatus should be compact. The layout of apparatus should be designed properly. Therefore, the reactor, catalyst, and reducing agent tank should be compact in the SCR system. To guarantee the endurance of the catalyst, back data are necessary. The degradation of catalyst and deposit due to diesel particulates and gas components in the exhaust gas should be considered in the design and development of the SCR system. For instance, the structure should be considered to replace the catalyst in the reactor, and thermal regeneration should be considered in the SCR downstream of a supercharger. The amount of reducing agent injection is determined by the NO_x concentration in the exhaust gas and the De-NO_x efficiency. An appropriate amount of reducing agent injection is estimated based on the amount of reduced NO_x. A feedback control that can be performed by measuring the NO_x amount at the SCR outlet is also effective and will be explained next.

5.7.2 Sensing technology for marine SCR and EGR control

The onboard measurements for the investigation and regulation of actual gas conditions are garnering attention. For automobiles in the United States of America and the European Union, to examine the exhaust gas level of the used vehicle, a real road running test using a gas measurement system (PEMS) is necessary for heavy-duty weighted vehicles such as trucks. In the European Union, even for small cars, as a requirement for new car model certification, from September 2017, the PEMS or a real road running exhaust gas test was introduced. In ships, the application for SCR of the PEMS should be considered as a matter of maintenances of the aftertreatment system. For onboard SCR feedback control, a measurable NO_x sensor meter has been developed. As an example, a NO_x analyzer for ship SCR control is explained based on the measurement principle, configuration, main specifications, and test results [36].

Fig. 5.7.1 shows the schematic of a ship's SCR. In the SCR, a catalyst is attached to the inside of an exhaust pipe, and a reducing agent of the urea

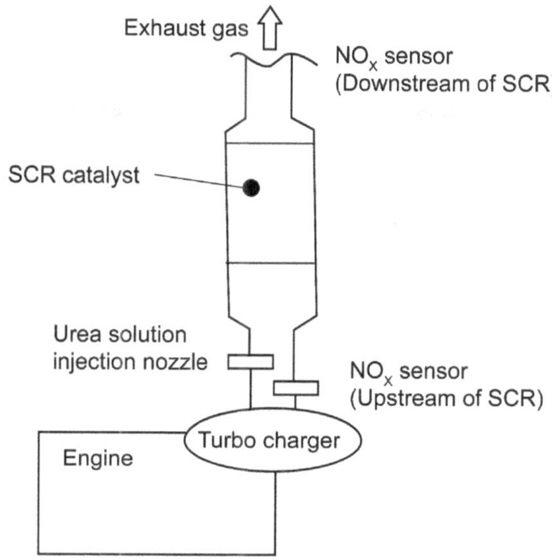

Figure 5.7.1 Schematic diagram of marine selective catalytic reduction (SCR) equipped with NO_x monitoring sensors [36].

solution can be injected by the nozzle. The injected urea solution is decomposed by the heat of the exhaust gas, and ammonia is induced. Thereafter, NO_x and ammonia contained in the exhaust gas is decomposed into harmless nitrogen and water by a selective catalytic reaction. When the amount of urea solution injected against the NO_x concentration contained in the exhaust gas is insufficient, sufficient De-NO_x cannot be performed. However, when excess urea water is injected, unreacted harmful ammonia is exhausted. To perform an efficient De-NO_x, the optimization of the urea injection amount is of utmost importance. Therefore, before the gas passes through the SCR catalyst, the NO_x sensor is attached to the pipe. The NO_x concentration is monitored in real time and is fed back to the SCR control such that the most appropriate urea water injection is performed to realize a highly efficient De-NO_x. Furthermore, a NO_x sensor is introduced downstream of the SCR to confirm the De-NO_x efficiency and possible NO_x emission. The details of the NO_x analyzer for the ship SCR control are explained. Because gas sampling systems are required for the conventional chemiluminescence method and nondispersive ultraviolet absorption method—based NO_x analyzer, it is difficult to miniaturize the analyzers, and it is not suitable for measurements onboard. Therefore, a zirconia solid electrolysis sensor for NO_x measurement has been developed, which can be inserted directly into the exhaust pipe.

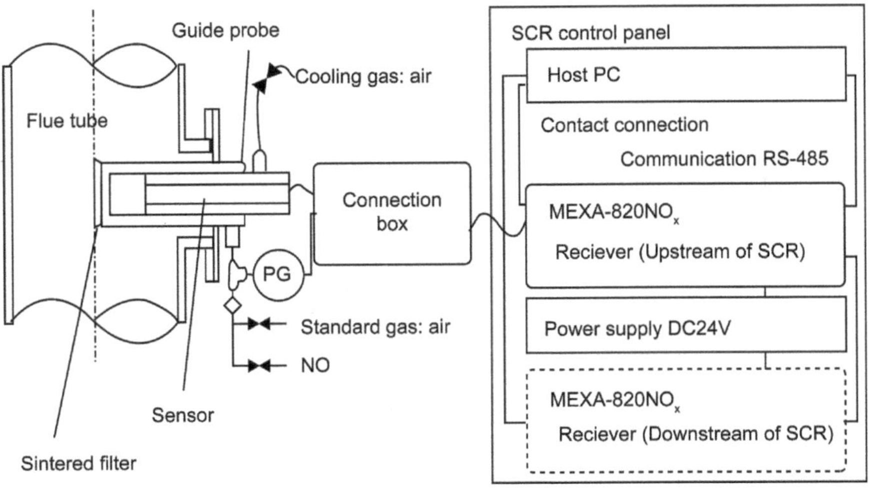

Figure 5.7.2 Device configuration diagram of NO_x monitoring system [36]. *SCR*, selective catalytic reduction.

Fig. 5.7.2 shows the device configuration diagram of the system (MEXA-820NO_x, Horiba Ltd., Japan). The sensor is inserted into the guide probe and placed in the exhaust pipe. It is inserted directly. The exhaust gas flowing through the exhaust pipe is guided by a pipe, reaching the sensor by diffusing from the sintered filter at the tip of the probe. Therefore, additional sampling parts are unnecessary. The unit captures the signal from the sensor and converts it to NO_x concentration in real time. The external input/output data transmission function of the RS-485 exists. The NO_x concentration is transmitted to the upper host PC to control the SCR system. Further for the sub engines equipped in the ships as several power generation engines, if MEXA-820NO_x is installed in the exhaust pipe of each engine, all the NO_x data in the exhaust gas of each engine can be collected simultaneously.

This NO_x sensing, monitoring, and control technology is developed for the marine SCR system. However, this NO_x sensing technology is also considered to be important for the optimal EGR operation in terms of the balance of output power and NO_x emission. This kind of NO_x sensing and control technology would be applied for the marine EGR system.

References

[1] K. Watanabe, High operation capable marine dual fuel engine with LNG, J. Jpn. Inst. Mar. Eng. 50 (6) (2015) 738–743.

[2] I. Ohashi, Dual-fuel marine engine (Highly reliable environmentally friendly engine), YANMAR Technical Review, 2015, 2018, https://www.yanmar.com/en_th/technology/technical_review/2015/0727_2.html.

[3] K. Hironaka, T. Miura, K. Watanabe, Y. Kurai, K. Yuki, Development of marine dual fuel engine "28AHX-DF", IHI Eng. Rev. 48 (2) (2015) 25–28.

[4] M. Okubo, Development of super-clean diesel engine and combustor using nonthermal plasma hybrid aftertreatment, J. Phys. Conf. Ser. 646 (1) (2015) 012051, total 6 pages.
[5] H.-J. Kim, B. Han, C.G. Woo, Y.-J. Kim, NOx removal performance of a wet reduction scrubber combined with oxidation by an indirect DBD plasma for semiconductor manufacturing industries, IEEE Trans. Ind. Appl. 54 (6) (2018) 6401–6407.
[6] H. Yamamoto, T. Kuroki, H. Fujishima, Y. Yamamoto, K. Yoshida, M. Okubo, Pilot-scale exhaust gas treatment for a glass manufacturing system using a plasma combined semi-dry chemical process, IEEE Trans. Ind. Appl. 53 (2) (2017) 1416–1423.
[7] W. Balachandran, N. Manivannan, R. Beleca, M. Abbod, Reduction of NO_x and PM in marine diesel engine exhaust gas using microwave plasma, J. Phys. Conf. Ser. 646 (2015) 012053.
[8] W. Balachandran, N. Manivannan, R. Beleca, M.F. Abbod, D. Brennen, N.S. Alozie, L.C. Ganippa, Nonthermal plasma system for marine diesel engine emission control, IEEE Trans. Ind. Appl. 52 (3) (2016) 2496–2505.
[9] M. Hołub, S. Kalisiak, T. Borkowski, J. Myśków, R. Brandenburg, The influence of direct non-thermal plasma treatment on particulate matter (PM) and NO_x in the exhaust of marine diesel engines, Pol. J. Environ. Stud. 19 (6) (2010) 1199–1211.
[10] K. Hanamoto, SO_x reduction technologies and development of dual fuel engine, J. Jpn. Inst. Mar. Eng. 50 (3) (2015) 336–340 (in Japanese).
[11] M. Adachi, H. Kosaka, T. Fukuda, S. Ohashi, K. Harumi, Economic analysis of trans-ocean LNG-fueled container ship, J. Mar. Sci. Technol. 19 (2014) 470–478.
[12] W. Butte, B. Heinzow, Pollutants in house bust as indicators of indoor contamination, in: G.W. Were (Ed.), Reviews of Environmental Contamination and Toxicology, vol. 175, Springer, New York, 2002, pp. 1–46.
[13] Ship & Ocean Foundation, Heisei 16 Nendo Senpaku karano VOC no Haishutsu Eikyo ni kansuru Chosa Houkokusyo (2004 Research Report on Influence of Volatile Organic Compounds (VOCs) Emission from Ships), 2005 (in Japanese).
[14] International Maritime Organization, MARPOL Consolidated Edition 2006, 2018, http://www.marpoltraining.com/MMSKOREAN/MARPOL/Annex_VI/r15.htm.
[15] Bureau of Environment Tokyo Metropolitan Government, Tokyo-to VOC Taisaku Gaido (Kojonai hen) (Guidebook for Measures on VOC in Tokyo [in Factory]), 2006 (in Japanese).
[16] The Ministry of the Environment, Taiki Osen Jyokyo (Air Pollution Situation), 2018 (in Japanese), http://www.env.go.jp/air/osen/joko_h26/rep07_h26.pdf, http://www.env.go.jp/air/osen/jokyo_h26/rep03_h26.pdf.
[17] M. Shima, $PM_{2.5}$ oyobi ozone ni yoru taiki osen to sono kenko eikyo (Effect of $PM_{2.5}$ and ozone on air pollution and human health), in: Distributed Document of Public Open Lecture Supported by Environment Research and Technology Development Fund of the Ministry of the Environment, Japan, No. 5-1456, March 24, 2017, (in Japanese)
[18] M. Shima, $PM_{2.5}$ oyobi ozone no kokyukikei eno eikyo (Effect of $PM_{2.5}$ and ozone on respiratory system), in: Distributed Document of Public Open Lecture Supported by Environment Research and Technology Development Fund of the Ministry of the Environment, Japan, No. 5-1456, March 24, 2017, (in Japanese)
[19] S. Yamazaki, M. Shima, Y. Yoda, F. Kurosaka, T. Isokawa, S. Shimizu, T. Ogawa, N. Kamiyoshi, K. Terada, J. Nishikawa, K. Hanaoka, T. Yamada, S. Matsuura, A. Hongo, I. Yamamoto, Association between chemical components of $PM_{2.5}$ and children's primary care night-time visits due to asthma attacks: a case-crossover study, Allergol. Int. 68 (3) (2019) 329–334. https://doi.org/10.1016/j.alit.2019.01.001.

[20] T. Sakurai, Senpaku wo taisyo to shita haisyustu kisei to nijiryushi noudo tono kankei (Relationship between emission regulation and secondary particle concentration by ships), in: PM2.5 Research Seminar "Influence of Domestic Sources" at Hyogo Prefectural Institute of Technology, Research Group on Chemical Characteristics of PM2.5, Japan Association of Aerosol Science and Technology, March 3, 2017, (in Japanese)
[21] T. Nagashima, Short-lived climate pollutants: SLCPs, December 28, 2012 (in Japanese), https://www.nies.go.jp/kanko/news/31/31-5/31-5-04.html.
[22] K. Takasaki, CO_2 reduction from main engine, J. Jpn. Inst. Mar. Eng. 50 (2) (2015) 198−201 (in Japanese).
[23] A. Azetsu, Regulation of GHG emissions and trend of countermeasures, J. Jpn. Inst. Mar. Eng. 51 (1) (2016) 50−53 (in Japanese).
[24] S. Izumi, Waste heat recovery and propulsion assist system for ships, J. Soc. Automot. Eng. Jpn. 70 (5) (2016) 63−68 (in Japanese).
[25] T. Takaishi, Approach to high efficiency marine low speed diesel engine, J. Jpn. Inst. Mar. Eng. 50 (2) (2015) 210−214 (in Japanese).
[26] K. Arai, J. Nagata, T. Uchida, S. Murakoshi, CO_2 reduction activities in a shipping company, J. Jpn. Inst. Mar. Eng. 50 (2) (2015) 244−246 (in Japanese).
[27] S. Nishio, GHG (greenhouse gas) reduction by biofuel use, J. Jpn. Inst. Mar. Eng. 51 (1) (2016) 61−65 (in Japanese).
[28] S. Nakao, R & D for marine engine using biofuel−application of refined bleached deodorized palm oil (RBDPO), J. Jpn. Inst. Mar. Eng. 50 (2) (2015) 66−71 (in Japanese).
[29] CISON PR Newswire, Exhaust systems market by after treatment device, fuel type, component, and region, aftermarket by component & after treatment device−global forecast to 2021, February 28, 2017, http://www.prnewswire.com/news-releases/exhaust-systems-market-by-after-treatment-device-fuel-type-component-and-region-aftermarket-by-component–after-treatment-device—global-forecast-to-2021-300415436.html.
[30] KFW IPEX-Bank, Will tightening regulations provide a silver lining for the merchant shippings sector? Maritime Industries, A Flash Analysis, January 2017. http://www.kfw-ipex-bank.de/International-financing/KfW-IPEX-Bank/Analyses-and-Views/Market-analyses/Maritime-Industries/.
[31] The Japanese Ministry of the Environment, Kankyo sangyo no shijyo kibo, Koyo kibo tou ni kansuru houkokusyo (Estimation of the market size of the environmental industries on a commission basis in fiscal year 2015), in: Report of Environmental Industry Market Size Research Meeting, March 2016 (in Japanese).
[32] Fuji Electric Co., Ltd, Shinseihin−Sennpaku muke haigasu jyouka souchi no hatsubai ni tsuite (New product of marine exhaust gas cleaning system is in market), April 8, 2018 (in Japanese), http://www.fujielectric.co.jp/about/news/detail/2014/20140408130012953.html.
[33] Shipbuilding Statistics, The Shipbuilders' Association of Japan, April 2017, http://www.sajn.or.jp/e/press/view/c6c7fdceed4cc2ecf535dcb294ab795ac7fbf6bc.
[34] T. Mizunuma, Kikan Tori Atsukai Taboo Syu (Taboo in Engine Operations), Seizan-do, 2006 (in Japanese).
[35] H. Sueshige, SO_x/NO_x kisei ni kannsuru sekkei kaihatsu doko (Trend of design and development for SO_x/NO_x regulation), in: Report by Maritime Innovation Japan Corporation, 2014 (in Japanese), http://iritokyo.co.jp/documents/mijac2014nov25.pdf.
[36] H. Naya, NO_x analyzer for marine SCR control: non-sampling NO_x analyzer using zirconia NO_x sensor, J. Jpn. Inst. Mar. Eng. 52 (3) (2017) 324−328 (in Japanese).

Appendix

A.1 Image of Sakai city in 16th century

People began living in the region around present-day Sakai City, Osaka Prefecture, Japan, approximately 10,000 years ago. In the 4th and 5th centuries, the Imperial Court was established, and more than 100 Emperor Tombs were constructed in Sakai City. "Sakai" means a boundary or border in Japanese; the area was named as such because it is located at the boundary of the three small prefectures that were established at the time. In the 12th to 14th centuries, Sakai was developed as a main shipping base of western Japan. At the end of the 15th century to the end of the 16th century, when domestic wars frequently occurred in Japan, Sakai developed significantly as an international trade port. Although Japan was not open to foreign countries in this era, Sakai prospered as a unique independent city known as "Saccai" and as an international trade base with countries such as China and Europe. An image of the historical prosperous Sakai in the 16–17th century was drawn and published in a book written by the Dutch missionary and scholar Arnoldus Montanus [1], as shown in Fig. A.1. Several ocean-going international ships are observed near the Sakai port.

After the 19th century, modernization progressed rapidly, including the development of modern industries, the expansions of the population and the city area, and increase in traffic, similar to other big cities in Japan. Although many ruins and prehistoric sites were lost in the bombings during World War II, Sakai continues to develop as an industrial city in Japan.

A.2 MARPOL 73/78 convention Annex I–VI

MARPOL 73/78 convention Annex I–VI [2,3] is explained. MARPOL 73/78 convention Annex I, which contains regulations for the prevention of oil pollution, entered into force on October 2, 1983. It covers prevention of oil pollution from operational measures and from accidental discharge into the sea; the 1992 amendments to Annex I made it mandatory for new oil tankers to have double hulls and include a phase-in schedule for existing tankers to fit double hulls, which was subsequently revised in 2001 and 2003.

Figure A.1 Sakai city imaged by Arnoldus Montanus (c.1625–83) [1].

MARPOL 73/78 convention Annex II, which contains regulations for the control of pollution by noxious liquid substances in bulk, entered into force on April 6, 1987. It details the discharge criteria and measures for the control of pollution by noxious liquid substances carried in bulk. Around 250 substances were evaluated and included in the list appended to the convention. The discharge of residues is allowed only to reception facilities on compliance with certain concentrations and conditions (which vary with the category of substances). In any case, no discharge of residues containing noxious substances is permitted within 12 miles of the nearest land.

MARPOL 73/78 convention Annex III, which covers the prevention of pollution by harmful substances carried by sea in packaged form, entered into force on July 1, 1992. It contains general requirements for the issuance of detailed standards on packing, marking, labeling, documentation, stowage, quantity limitations, exceptions, and notifications. For the purpose of this Annex, The harmful substances are substances identified as marine pollutants under the International Maritime Dangerous Goods Code or those that meet the criteria in the Appendix of Annex III.

MARPOL 73/78 convention Annex IV, which covers the prevention of pollution by sewage from ships, entered into force on September 27, 2003. It contains requirements to control pollution of the sea by ship sewage. The discharge of sewage into the sea is prohibited, except when the ship has an approved sewage treatment plant in operation or when the ship discharges comminuted and disinfected sewage using an approved system at a distance of more than three nautical miles from the nearest land. Sewage that is not comminuted or disinfected must be discharged at a distance of more than 12 nautical miles from the nearest land.

MARPOL 73/78 convention Annex V, which covers the prevention of pollution by garbage from ships, entered into force on December 31, 1988. It deals with different types of garbage and specifies the distances from land and the manner in which they may be disposed of. The most important feature of Annex V is the complete ban on the disposal of all forms of plastic into the sea.

In 1997, a protocol was adopted to amend the convention, and Annex VI was newly added to Annex I–V. MARPOL 73/78 convention Annex VI, which covers the prevention of air pollution from ships, entered into force on May 19, 2005. It sets limits on emissions of sulfur oxide (SO_x), nitrogen oxide (NO_x), volatile organic compounds (VOCs), and shipboard incineration from ship exhausts. It also prohibits deliberate emissions of ozone-depleting substances such as chlorofluorocarbons (CFCs). Designated Emission Control Areas (ECAs) set more stringent standards for SO_x, NO_x, and PM. An amendment adopted in 2011 covers mandatory technical and operational energy efficiency measures aimed at reducing emissions of greenhouse gases. In Annex VI, Regulation 1–4 mention general information of Annex VI such as application of Annex VI, definition, general exceptions of this application, and allowance of alternative way equivalent to Annex VI. Regulation 5–11 mention survey, certification, and means of control, such as procedure of survey, issue or endorsement of certificate, form of certificate, duration and validity of certificate, control on operational requirements in a port or an offshore terminal, and detection of violations and enforcement. The details of Regulation 12–23 are as follows:

1. Target substances

   Ozone-depleting substances such as CFCs, NO_x, SO_x, VOCs, and shipboard incineration substances.

2. Application

   All ships including fixed and floating platforms are subject to regulation. Ships of 400 GT class and more are subject to inspection.
3. Regulation on ozone-depleting substances such as halons and CFCs (Regulation 12)

   Any deliberate emissions of ozone-depleting substances are prohibited, except minimal releases associated with the recapture or recycling of an ozone-depleting substance. When ozone-depleting substances and/or equipment containing such substances are removed from ships, these must be delivered to the appropriate reception facilities. New installations that contain ozone-depleting substances are prohibited on all ships except new installations containing hydrochlorofluorocarbons, which are permitted until January 1, 2020.
4. Regulation on nitrogen oxide (NO_x) (Regulation 13)

   This regulation applies to diesel engines with a power output of more than 130 kW, installed in ships that were constructed on or after January 1, 2000. This regulation also applies to diesel engines with a power output of more than 130 kW, which underwent a major conversion on or after January 1, 2000. This regulation also does not apply to emergency diesel engines; engines installed in lifeboats, domestic coastal trading ships, and any device or equipment intended for use only in case of emergency; and engines installed in ships that only travel within waters subject to alternative NO_x control measures approved by the administration. It is noted that the administration means the government which has the authority of the ship. NO_x emission from the engine, which is calculated as the total weighted emission of NO_2, should be within the limits set by the International Maritime Organization (IMO).
5. Regulation on sulfur oxide (SO_x) (Regulation 14)

   As a general requirement, the sulfur content of any fuel oil used onboard ships should not exceed 4.5 mass%. As for requirements within SO_x Emission Control Areas (ECAs for SO_x) such as the Baltic Sea area, the North Sea area, the North American area, and the Caribbean Sea area at least one of the following conditions should be met: (1) The sulfur content of fuel oil used on board ships in an ECA for SO_x should not exceed 1.5 mass%. (2) An exhaust gas cleaning system, which is approved by the administration taking into account guidelines to be developed by the IMO, should be adopted to reduce the total SO_x emission from ships, including both auxiliary and main propulsion engines, to 6.0 g/kWh or less. (3) Any other technological method that is verifiable and enforceable to limit SO_x emissions to a level equivalent to that in (2) should be applied. These three methods should be approved by the administration taking into account guidelines to be developed by the IMO.

6. Regulation on volatile organic compounds (VOCs) (Regulation 15)

   If VOC emissions from tankers are to be regulated in ports or terminals under the jurisdiction of a contracting country to MARPOL 73/78 convention, they are regulated in accordance with the provisions of this regulation. This regulation is applied only to gas carriers when the type of loading and containment system allows safe retention of nonmethane VOCs on board or their safe return ashore. A contracting country to MARPOL 73/78 convention, which designates the ports or terminals under its jurisdiction where VOC emissions are to be regulated, should submit a notification to the IMO. This notification should include information on the size of tankers to be controlled, cargoes requiring vapor emission control systems, and the effective date of such control. The notification should be submitted at least 6 months before the effective date. A vapor emission control system, which is approved by that government taking into account the safety standards, should be installed in ports or terminals where VOC emissions from tankers are to be regulated. The vapor emission control system should be operated during the loading and unloading of tankers such as chemical tankers.

7. Regulation on shipboard incineration from ship exhausts (Regulation 16)

   Shipboard incineration shall be allowed only in a shipboard incinerator. Each incinerator installed onboard a ship on or after January 1, 2000 should meet the requirements for oxygen concentration in an incinerator and temperature limits in emission, which are regulated by the IMO. Each incinerator shall be approved by the administration taking into account the standard specifications for shipboard incinerators developed by the IMO. The administration can grant exclusion from this application to any incinerator installed on board a ship before the date of entry into force of the protocol of 1997. Shipboard incineration of the following substances is prohibited: (1) cargo residues regulated under Annex I, II, and III of MARPOL 73/78 convention and related contaminated packing materials; (2) refined petroleum products containing halogen compounds; (3) garbage, as defined in

Annex V of MARPOL 73/78 convention, containing more traces of heavy metals; and (4) polychlorinated biphenyls (PCBs)

8. Reception facilities (Regulation 17)

The government of each contracting country to MARPOL 73/78 convention should undertake steps to ensure the provision of adequate facilities to meet the needs of ships using its repair ports for the receipt of ozone-depleting substances and equipment containing such substances on removal from ships and exhaust gas cleaning residues from approved exhaust gas cleaning systems.

9. Fuel oil quality (Regulation 18)

Fuel oil delivered to and used on board ships for combustion purposes should meet the following requirements. The fuel oil should be hydrocarbons blends derived from petroleum refining. The fuel oil with small amounts of additives intended to improve some aspects of performance is permitted. However, the fuel oil should be free of inorganic acid. The fuel oil should not contain any of the following substances or chemical waste: (1) substances that jeopardize the safety of ships or adversely affect the performance of the machinery; (2) substances that are harmful to personnel; or (3) substances that contribute to overall additional air pollution. For every ship of 400 GT and above and every fixed and floating drilling rig and other platforms, details of the fuel oil for combustion purposes delivered to and used on board should be recorded by means of a bunker delivery note containing information such as marine fuel oil supplier, product name, quantity in metric tons, density at 15°C in kg/m^3, sulfur content in mass%, and ship and port information. The bunker delivery note should be kept aboard the ship in such a place as to be readily available for inspection at all reasonable times. It should be retained for a period of 3 years after the fuel oil has been delivered on board. The competent authority of the government of a contracting country to MARPOL 73/78 convention can inspect the bunker delivery notes on board any ship while the ship is in its port or offshore terminal. The bunker delivery note should be accompanied by a representative sample of the fuel oil delivered within a period of no less than 12 months, taking into account the guidelines to be developed by the IMO. Local fuel oil suppliers should provide the bunker delivery note with a sample and retain a copy of the bunker delivery note for at least 3 years for inspection.

10. Regulation on carbon dioxide (CO_2) (Regulation 19–23)

The amendments to the MARPOL 73/78 were adopted in July 2011. Regulations on energy efficiency for ships were added to MARPOL Annex VI to make mandatory the Energy Efficiency Design Index (EEDI). EEDI is an index of mass of CO_2 emission from ship in 1 mile transport of 1 ton of cargo. Regulation 19 states that it is applied to all new ships of 400 gross tonnage and above, which are engaged in international water. Regulation 20 mentions the attained EEDI, that is, the actual EEDI, which is calculated by EEDI formula according to type of ship. The attained EEDI should be verified by the government in own country, or an authorized organization. Regulation 21 mentions the required EEDI as shown in Table 5.5.1. The regulation defines the four reduction phases, i.e., phases 0 for 2013–2014, phase 1 for 2015–2019, phase 2 for 2020–2024, and phase 3 for 2025–. The required EEDI reduction rates are 0, −10%, −20%, and −30% for phases 0–3. Regulation 22 mentions the Ship Energy Efficiency Management Plan (SEEMP). In the SEEMP, measures and steps for improving the energy efficiency of ships to reduce CO_2 emission should be described. EEDI is also included in SEEMP for required ships. SEEPM is required for all ships to keep on board. Regulation 23 mentions promotion of technical co-operation and technology transfer. Developing countries may face both of technological and financial difficulties as a result of the energy efficiency regulations. To reduce these difficulties, Regulation 23 requests that the countries promote and provide technical co-operation, technical assistance, and technology transfer in co-operation with the organization and other international bodies.

A.3 Definition of plasma temperature

The definition method for electron temperature of plasma is explained. Generally, the temperature can be defined only when the kinetic energy of each particle follows the Maxwell distribution. The electron temperature for electrons, the ion temperature for ions, and the gas temperature or simply the temperature for heavy particles can be defined. Electrons and heavy particles (neutrals and ions) have respective kinetic energies and continuously collide with each other. Most collisions are considered completely elastic and can be treated based on the theory of ideal gas. Therefore, particles as a whole follow a simple law, although individual particles move randomly. The velocity of a particle follows the Maxwell distribution law after many collisions occur.

$$f(u_x) = \left(\frac{m_e}{2\pi kT_e}\right)^{1/2} \exp\left(-\frac{m_e u_x^2}{2kT_e}\right) \quad \text{(A.1)}$$

$$f(u_y) = \left(\frac{m_e}{2\pi kT_e}\right)^{1/2} \exp\left(-\frac{m_e u_y^2}{2kT_e}\right) \quad \text{(A.2)}$$

$$f(u_z) = \left(\frac{m_e}{2\pi kT_e}\right)^{1/2} \exp\left(-\frac{m_e u_z^2}{2kT_e}\right) \quad \text{(A.3)}$$

$$F(u) = 4\pi\left(\frac{m_e}{2\pi kT_e}\right)^{3/2} u^2 \exp\left(-\frac{m_e u^2}{2kT_e}\right) \quad \text{(A.4)}$$

The probability density functions of particle velocities in Eqs. (A.1)–(A.3) give the cumulative distribution function $F(u)$ of particle velocity $u = (u_x + u_y + u_z)^{1/2}$ in Eq. (A.4), defined in the velocity space (u_x, u_y, u_z). Here, electrons are considered as an example, and the same treatment holds for ions by replacing the mass and temperature of the electrons with those of the ions.

The electron temperature T_e is defined as

$$u_p = \left(\frac{2kT_e}{m_e}\right)^{1/2} \quad \text{or} \quad \frac{1}{2}m_e u_p^2 = kT_e \quad \text{(A.5)}$$

where the electron exists most probably. This velocity is the most probable velocity. Eq. (A.5) defines the electron temperature.

References

[1] A. Montanus, The Citty SACCAI, Atlas Japannensis, published in 1670, International Research Center for Japanese Studies, http://www.nichibun.ac.jp

[2] International Maritime Organization (IMO), International Convention for the Prevention of Pollution from Ships (MARPOL), 2018, http://www.imo.org/en/About/Conventions/ListOfConventions/Pages/International-Convention-for-the-Prevention-of-Pollution-from-Ships-(MARPOL).aspx.

[3] International Maritime Organization (IMO), Annex VI —regulations for the Prevention of Air Pollution from Ships, 2018, http://www.marpoltraining.com/MMSKOREAN/MARPOL/Annex_VI/index.htm.

Index

Note: 'Page numbers followed by "t" indicate tables, "f" indicate figures'.

I

J

K

L

M

N

O

P

Kolibri 360 GmbH
Ferdinand-Jühlke-Straße 7
99095 Erfurt, Deutschland
produktsicherheit@kolibri360.de